A DAY AT THE BOTTOM OF THE SEA

A DAY AT THE

BOTTOM OF THE SEA

by James Parks

Crane-Russak   New York

A Day at the Bottom of the Sea
Published in the United States by
Crane, Russak & Company, Inc.
347 Madison Avenue
New York, New York 10017

Copyright © 1977 Crane, Russak & Company, Inc.

ISBN 0-8448-1045-2 √
LC 76-46209

Designed by Meri Shardin

Printed in the United States of America

*To Miss Carmie Wolfe, my English teacher
through high school, who somehow knew that one day I
would write a book.*

Contents

Preface

THIS IS THE STORY OF ONE DAY AT THE BOTTOM OF THE SEA. THE day was October 21, 1970. The occasion was a dive on a research submarine, an experience that few people can have at first hand. This is the way it happened to me.

Here, in these pages, is some of the awe and wonder that comes with such an adventure, and some of the excitement of scientific research into the unknown. The dive was the first of a series, as part of a research program conducted by Lehigh University and funded by The National Sea Grant Office of the United States Department of Commerce, under the National Oceanographic and Atmospheric Administration (NOAA). This was one of the first Sea Grant projects to be undertaken by a university–industry partnership. Lehigh's industrial partner on this project was Lockheed Missiles and Space Company, Inc., which designed, built, and operated the research submersible *Deep Quest.*

For their help in making this experience possible, I wish to thank Bob Abel, Hal Goodwin, and Bob Wildman of the Sea Grant office; Professors Adrian Richards, Terry Hirst, and Ted Terry of Lehigh University, and the several Lehigh University graduate students associated with the project; and Tony Inderbitzen, Frank Simpson, and Bill Crenshaw of the Lockheed Ocean Laboratory. Special thanks go to Larry Shumaker and Donald Saner, pilot and navigator of *Deep Quest* on this dive; and to the skipper, officers, and crew of the submarine's mother ship *Transquest.*

Walnutport, Pennsylvania
January 22, 1977

Glimpses Through a Window to Another World

ALTHOUGH I WAS IN EXTREME PHYSICAL DISCOMFORT, AT TIMES almost to the point of pain, I was so fascinated with what I was seeing that I didn't want to quit. I was lying on my stomach on icy cold bare steel that was curved the wrong way for comfort. The space was too narrow and too short. My knees were jammed against a metal wall, an unyielding metal edge cut into my shins, and I couldn't shift into a more comfortable position. Worst of all, my head was bent forward in an awkward position, and in order to see I had to look over the tops of my glasses.

I was peering through a window of very thick glass, six inches in diameter, at a world I had never seen before. Strange and weird objects moved through my narrow line of vision. Nothing was familiar. The scale and size of the objects could not be guessed. The sensation of speed was unnerving— objects flashed across the brilliantly lighted scene too quickly to be identified. What was that? A shiny metallic cylinder lying on its side, a triangular hole in one end—a beer can!

Suddenly the strain of the awkward position was too much for me. My neck muscles hurt. I was cold. And I had a cramp in one leg. Cautiously, with head held low, I backed and twisted until I was in a kneeling position and could raise my head without bumping it. When I turned around, two familiar faces were grinning at me—my shipmates and co-workers on this cruise. I had had my first turn at the visual observation port in a small submarine at the bottom of the sea more than 4,000 feet down.

One of my shipmates reached out his hand and helped me to my feet. Then we changed places, and he bent over to worm his way into position at the viewport. I rubbed my stiff neck—even here I had no room to stretch my aching muscles —and looked around at our crowded vehicle. I wondered again why in the world I had volunteered for such an uncomfortable and hazardous trip. But all the while I knew I was enjoying every minute of the adventure and wouldn't have missed it for anything. Sometimes science can be exciting.

What was I doing in a submarine at the bottom of the sea? Why was I given the chance to go on the trip? What was the purpose of this submarine and of this dive? What does it feel like to be in a submarine? What does the bottom of the sea look like? Read on and share in my day at the bottom of the sea.

General Preparations

My transcontinental jet airliner arrived at San Diego International Airport late at night. After retrieving my suitcase in the baggage claim area, I went to the front door of the Terminal Building. My destination was only a few blocks away, so I ignored the waiting taxis and walked into the balmy autumn night. It was about midnight when I entered the gates of the Lockheed Ocean Laboratory on Harbor Island. After I had identified myself and my purpose, the watchman at the gate directed me to the dock behind the main building. As I came around the corner I had my first look at the research submarine named *Deep Quest*, on which I was to spend a day at the bottom of the sea.

For months I had studied pictures and drawings of *Deep Quest*. Now, at last, I would be able to get a close look at it and even touch it.

Moored at the dock was a small ship, lights blazing, dozens of men scurrying about. Were there really dozens of them? I didn't stop to count. On the afterdeck of the ship I saw a large white object, shaped like a teardrop, glistening under the bright lights. For a moment it looked like a sleeping white whale, complete with eyes and a mouth. Then those features became camera lenses and lights and instrument racks. I saw men near the conning tower on top and other men making adjustments to various pieces of equipment. *Deep Quest* stopped looking like a whale and became a machine again.

As I stood on the dock and looked up at this strange vessel, I had momentary doubts about setting out on this adventure.

The research submarine Deep Quest *in its hangar at The Lockheed Ocean Laboratory, appearing much as it did when I first saw it on the floodlit deck of its mother ship* Transquest. *It really looks like a beached white whale.*

Lockheed Ocean Laboratory on Harbor Island, San Diego. The mother ship Transquest *is moored at the dock. The submersible* Deep Quest *can barely be seen in the hangar, the large building at the upper right.*

Yet this was something I had long wanted to do and I shook off my doubts. As a soft-rock geologist, I studied marine sedimentary rocks that had formed untold thousands of years ago in seas that once covered the land. I knew in theory the natural processes that formed these rocks—sandstones, limestones, shales—but I was short on first-hand observation of some kinds of rocks in the process of being formed.

I had studied sandy beaches in Texas, California, and New Jersey, where quartz sand grains were being rounded by wave and surf action. At some future time, perhaps these sandy sediments will be buried by other sediments, and will be lithified or hardened into sandstone by the precipitation of minerals from natural chemicals dissolved in water.

Off the southern tip of Florida, in sheltered waters behind the Florida Keys and near Andros Island in the Bahamas, I had waded in limy muds composed of the mineral calcium carbonate that would someday become limestone. I had snorkeled and scuba-dived around living coral reefs on the seaward side of the Florida Keys, in the Bahamas and near Bermuda, to see other types of limestone being formed.

But the most abundant sedimentary rock is shale. And shale usually forms from mud deposited in deeper water, beyond the depths where the amateur scuba diver can go. My experience with shale muds was limited to small samples taken from the sea floor by apparatus lowered to the bottom on cables from surface oceanographic ships. It is a truism that one picture can portray an unfamiliar object in a way that not even a thousand words can do. Just so, a few hours of looking at the sea floor could well be worth a thousand small samples of the bottom muds in making me understand and appreciate the sediment-forming processes. Tomorrow I would go to the bottom myself and look at clay muds on the deep sea floor.

It had been a long day, and tomorrow would seem longer still. I had better get some sleep. Stepping over and around a maze of thick black electrical cables and hoses that stretched from the dock to the ship, I approached the ship's gangplank and climbed aboard. I was totally unfamiliar with the ship, but I knew that scientists and crew generally have separate living areas on oceanographic research ships. So I asked a workman how to get to the scientific quarters.

Following his directions, I climbed an iron ladder to an upper deck behind the wheelhouse and smokestack. There I saw two deckhouses, shaped like great big boxes, without windows. The first one I entered was the scientific instrument shop, where a couple of fellow scientists were going over plans for tomorrow. They welcomed me aboard, told me that I was scheduled for the first dive and that the ship would get underway soon, and suggested I find a bunk in the deckhouse next door. In it I soon spotted an empty upper bunk with my name taped to the frame, and gratefully I went to bed and to sleep.

The midship deckhouses installed as scientific workshops and dormitories crowd and clutter up the decks of Transquest.

During construction of Deep Quest *in 1967, the bispherical pressure hull is lowered into place within the outer hull. Note the upper access hatch in the rear pressure sphere. The rear-facing open hatch on the side of the rear sphere was designed to be connected to another set of pressure hulls for diver lockout and deep sea rescue work. These additional spheres were never installed.*

I had come all the way from a university on the East Coast to take part in a research dive on the *Deep Quest*. The dive was part of a project sponsored by the National Sea Grant Program to study the engineering properties of the deep ocean floor. Other scientists and engineers had been in San Diego for several days, installing and testing our specially designed equipment.

Between 0200 hours (2 A.M.) and 0300 hours (3 A.M.), while I slept, the hoses and electrical cables were disconnected, about half the workmen went ashore, and the ship put to sea. We were aboard the Research Vessel *Transquest*, an odd-looking ship with an open stern well. Boxlike deckhouses had been set down on the upper deck like afterthoughts.

The R/V *Transquest* is the mother ship—or surface support

vessel—for an even more unusual craft, Lockheed's research submersible *Deep Quest,* one of the earliest deep-diving submarines to be built by a private corporation for research in the deeper parts of the oceans.

At its top speed of about 4 knots, which is less than 5 miles per hour, the 108-foot ship carrying the 50-ton, 40-foot submarine proceeded to a position about 20 miles offshore, over a broad deep valley on the sea floor known as the San Diego Trough. The captain of *Transquest* could tell when we had reached our destination by taking compass bearings on lighthouses that were barely visible on the horizon.

For ordinary navigation on the open sea, knowing your location within a mile or so is usually good enough. For oceanographic research, we had to be a lot more precise. For example, we wanted to come back to the exact spot several times in the course of several years in order to complete a detailed survey that would take longer than the *Transquest* could stay at sea.

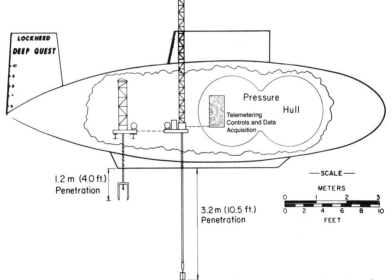

Schematic longitudinal cross-section of Deep Quest *showing location of bottom-testing probes and drive towers. Note that the submarine could have been considerably shorter and less expensive if it had not been planned to accommodate additional pressure spheres.* Deep Quest *was designed as an experimental prototype vessel rather than as a single-purpose work boat.*

Deep Quest *on the elevator in the stern of the mother ship.*

Nose of Deep Quest. *Screened oval opening behind coring device is intake-exhaust for vertical thruster. To right of coring device is bident gamma ray probe, temporarily attached to a manipulator arm. Cylindrical covering on bottom of right arm of bident is shield over the radioactive source.*

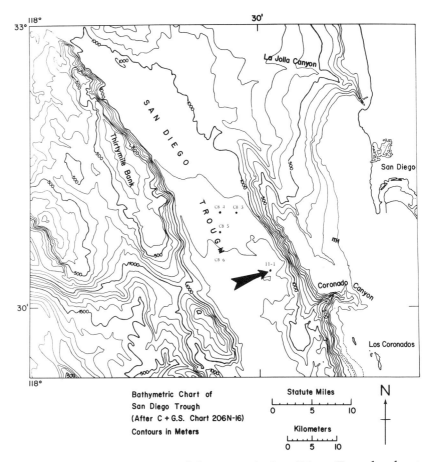

Bathymetric contour map of dive area in San Diego Trough, about 20 miles southwest of San Diego. Arrow points to dive site.

Our test area was marked with a semi-permanent marker on the bottom. But there was a limit to the amount of time we could stay down on each dive. And it would be a waste of valuable time to hunt around for the marker if we were to start as far as a mile or so away. Ideally, the sub ought to be launched almost directly above the bottom marker. This required unusually accurate navigation and pinpointing the exact location.

By far the best navigation and positioning system uses radio signals from special orbiting space satellites. A small shipboard electronic computer is then used to calculate the ship's position in latitude and longitude, which can then be plotted on a chart. Satellite navigation can routinely come within less than a hundred meters of the target, which is more than adequate for almost every purpose.

Nearly every large oceangoing research vessel now has a satellite navigation system, but on smaller ships like *Transquest*, which usually operate only in coastal waters, such a system costs too much.

Most of the dives made by *Deep Quest* have been fairly close to shore. This is because the southern California continental shelf is so narrow that deep water can be found in troughs and basins only a few miles offshore.

When the ship is within sight of the coast, taking visual bearings on landmarks or lights is the most accurate way to find the position of a ship on a map or chart. For those readers who would like to know more about triangulation with visual

Bird's-eye view of Transquest *and* Deep Quest *at sea.*

bearings, the Appendix at the back of the book explains the method in more detail.

Once a ship is on station, the captain wants to keep it at that location. Yet wind and ocean currents will tend to drift the ship away from the desired position. It is impractical to anchor in water deeper than about 100 meters (330 feet). What is more, on a mission such as that of the *Transquest*, anchor lines would interfere with the diving of the submarine. So the captain keeps the ship on station by what is known as dynamic positioning. This involves frequent checking of bearings to the onshore landmarks, as well as occasional repositioning of the ship. Twin diesel electrical generators were used in the dynamic positioning. The generators power two highly maneuverable 100 horsepower electric drive propulsion units.

During the night, while I was still asleep, the *Transquest* reached the dive site.

Dive Preparations

IT WAS STILL DARK WHEN SOMETHING AWAKENED ME. THE STEADY pulsating beat of the main engines had slowed to a low hum, and other noises intruded. Sounds of human activity— footsteps on iron ladders, a clang of metal on metal, the murmur of voices. A glance at my luminescent wristwatch dial told me the time: 0630 hours. We had reached our dive location, and preparations for the dive were beginning.

Fumbling for my clothes in the dark of this windowless, unfamiliar surrounding, I dressed hurriedly. Then I made my way to the crack of light around the door, opened it, and found a light switch. By the dim overhead light I could see four double bunks, each with its own small reading light. So as not to disturb the two sleepers, who had probably been up all night, I turned on my bedlamp, switched off the overhead light, then finished dressing. Carrying my toilet articles and a towel that had been draped over the foot of my bunk, I went out on deck.

The sun was up and the sea was nearly calm. Of the two iron ladders in front of me, one went up to what was obviously the bridge and captain's quarters, so I climbed down the other ladder to the main deck. Remembering some navy slang, I asked for directions to the "head," and having found it I washed and dressed. After returning my toilet kit and towel to my bunk, I followed my nose to breakfast.

I was so eager to see what was going on around the submarine that I intended to eat a light breakfast and pass up much of the plentiful and varied food always available on an oceanographic ship. Several men had arrived earlier and were already eating heartily. On the table were containers of juice, milk, dry cereal, and sweet rolls. The cook, at his stove in the galley on the other side of the small room, asked me what else I wanted. He was obviously frying eggs. I asked for two, sunny side up, with bacon. I drew a mug of coffee from the large urn, then joined the other at the table.

After saying "Morning" to the one man I knew, I introduced myself to the others. Meals on oceanographic ships are generally informal. There are always more people on board than can eat at one sitting in the usually small eating area. The hours of serving are posted on a bulletin board, and generally the start of serving is announced by a bell or chimes. Seating is first come, first served, and if you find the table full, you just wait around outside until someone leaves.

Because of the need to serve more people than can be seated at once, meals are eaten fast, there is little conversation, and no one lingers over coffee. Particularly at breakfast there is much coming and going. At other meals, usually served family style from bowls and platters on the table, all the people in the first sitting have to finish and leave the eating area before the second shift is seated and then served. Each person clears his own dishes from the table, scrapes any remains into a garbage pail, and stacks his dishes and eating utensils in the sink.

I had eaten a substantial breakfast. Then I left the table, making way for the late comers. Now, carrying my second mug of coffee, I paused at the bulletin board to check the meal hours and to read the announcements and instructions. Then I walked aft to the rear of the ship, where the submersible *Deep Quest* sat on a platform at deck level, and watched the crew get ready for the launch.

In a sense, there were several crews on board the *Transquest.* One small crew operated the mother ship in shifts around the clock; a larger crew maintained and serviced the submersible, launched it, and recovered it at the end of its dive; another small crew operated the submersible; and a scientific crew maintained the scientific instruments and made the observations and measurements that were the primary purpose of the cruise and dives.

In actual practice, these crews were not separate and distinct—everyone helped out as needed. There were probably more than two dozen people on board, although I never did get an accurate count. Most are highly skilled electricians, machinists, engineers, electronics and hydraulics experts, scuba divers, pilots for the submersible, and scientists.

Before each dive of *Deep Quest*, everything is checked out as thoroughly as it would be for a launch of a moon rocket. Checkout procedures take hours. Every component of every system is tested to make sure it will operate properly when and if it is needed. There are many complex systems—life support, communication, propulsion, navigation, electrical power, lighting—for the basic operation of the sub, systems for returning to the surface in the event of an emergency, and scientific instrument systems for carrying out the mission of the dive. Every system has at least one backup system in case of power failure or other malfunction.

One important difference between this submarine and a moon rocket is that the sub is used again and again—well over 100 dives had been logged at this time—and many components can begin to show the effects of aging. Between cruises when the *Deep Quest* is in port, it is thoroughly overhauled and anything showing wear is replaced. At sea between dives the checkout is also thorough, and an extensive supply of spare parts is kept on board the *Transquest* for replacements.

The environment of the deep oceans is at least as hostile to humans as is the environment of outer space. In either surrounding, direct exposure to the environment for only a few seconds means death. However, coming back from the bottom of the sea in case of minor trouble is a little easier than coming back from outer space. A few ballast weights can be dropped and the natural buoyancy of the sub will bring it back to the surface.

During the checkout period, only the pilot and the

navigator were in the sub, watching their instrument panel. A temporary telephone line connected them to the dive master, the man on board *Transquest* who was in charge of all shipboard systems related to the sub. Some checkout procedures were obvious to those of us who were watching. The pilot would turn on a switch and talk into the microphone hung around his neck, saying something like "Starboard forward floodlight on." The light would come on, the divemaster would observe it, and he would in turn talk into his telephone mouthpiece. "Starboard forward floodlight on—check." When the pilot heard this he would turn off that switch, tell the divemaster he had done so, and go on to the next item on his list. In this way they went through all lights, drive mechanisms, movements of control surfaces, and so on.

Other checkout procedures were not obvious to me because there was no visible or external reaction. After some components were checked, a safety device such as a lens cap or a block or pin to prevent further movement was attached: this was to prevent damage to that component during the launching operation. A large, brightly colored tag was attached to each safety device so it could be easily located and removed by scuba divers after the sub was in the water.

When the divemaster and the pilot were both satisfied that all was ready, the pilot and navigator climbed out and stood on the sub's deck around the conning tower. At a little after 0900 hours, the divemaster indicated that those of us who were scheduled for this dive could go on board the *Deep Quest*. Normally the sub carries four people: pilot, copilot–navigator, and two scientific observers. On this dive I was to be the fifth person, a supernumerary (extra) observer.

The three of us went to the rear of the mother ship, climbed a ladder, and walked across a small gangplank to the deck of the sub. We were told that we were not to enter the sub until after it was launched. Although all precautions were taken, launching and recovery of the sub are probably the most dangerous parts of the whole dive operation.

Launching and recovery can be accomplished only in a

fairly calm sea. When waves are higher than 3 or 4 feet, the motion of the ship (roll, pitch, and heave) is too large relative to the motion of the submersible, and the sub could bang against the side of the ship with an almighty thump.

The sub sits on an elevator platform that can be lowered into an open stern well of the ship. The rear of the ship is U-shaped as viewed from above, with two side extensions of the hull around a space that is open to the sea at the rear. In this stern well is the elevator platform, about 31 feet long and 23 feet wide, operated by hydraulic power and capable of lifting 60 tons.

During launching, the platform is lowered to about 12 feet below sea level. Since the *Deep Quest* has a draft of 8 feet 7 inches when floating, it floats free of the platform and can be backed out of the stern well under its own power. Several men stand on each walkway while the launch is in progress. Each of them holds one end of a heavy line that is attached to the sub. Together, they keep the sub centered in the stern well.

When all was ready on the *Deep Quest*, the pilot signaled to the dive master that launching could begin.

Launching the Submersible

AT A SIGNAL FROM THE DIVE MASTER, BOTH BY HAND AND by means of the telephone handsets, the engineer who was operating the elevator shifted his control lever. Slowly the elevator platform inched its way down, lowering the sub into the sea. As the sub descended the line handlers gradually paid out additional line. But they were careful to keep their end snubbed around a stanchion on the ship, so as to restrain the sub from moving sideways when it floated free of the platform. The sub's response to wave action is different from that of the larger and longer ship, so we could feel the change in motion of the sub as soon as she floated free.

While the launch was going on, we stood outside on the deck around the conning tower. The hatch leading to the inside of the sub was kept closed as a safety precaution. The procedures for launching submersibles are the result of trying to keep difficulties from developing, as well as of learning from the mistakes of others. A failure to anticipate the consequences of leaving the hatch open during launching had contributed to the sinking of the submersible *Alvin* on the East Coast.

The *Alvin*, a smaller research submarine operated by the Woods Hole Oceanographic Institution, was launched on a platform supported by four cables and a crane. As he had done on dozens of previous launchings, the pilot stood in the open hatchway of *Alvin* to direct the launching. The other crew member was inside the sub.

Shortly after the lowering began, one of the four cables

Deep Quest has just been lowered into the water in the open stern well of Transquest. The six lines (ropes) holding the sub in the center of the well will be slackened as the sub backs out under its own power, and then the men on the sub will cast off these lines.

broke, the platform tilted slightly, and the sub slid off and dropped several feet into the sea. The momentum of the fall carried the sub beneath the surface, and seawater flowed through the open hatchway. The *Alvin* bobbed back to the surface, allowing the pilot and the observer to scramble to safety through the open hatch. But the sub had lost too much buoyancy, and it sank in several thousand feet of water. Nearly a year went by before the *Alvin* could be recovered from the sea floor. Then, after a thorough checkout and minor repairs, it was put back into service. But never again was the hatch left open on any of the submersibles during the launch.

The deck we stood on was part of the outer hull of *Deep Quest*, made of fairly thin sheet aluminum and shaped like a streamlined teardrop. This outer hull was purposely not made watertight. There were holes that allowed seawater to flood

Beginning of launch operation. Elevator has lowered Deep Quest *into water, but lines are still attached to side walkways of Trans-quest.*

Deep Quest *has backed out of stern well. Scuba divers have come out in* Boston Whaler *(to left of sub) to make final checks and to remove safety chocks and lens caps.*

the outer hull so as to equalize the pressure on the inside and outside.

Soon after we floated free, the pilot of *Deep Quest* started the battery-powered electric thrusters. Then, using a hand-held control box, he began backing the sub slowly out of the stern well of the mother ship into the open sea. Although the mother ship itself was not moving, it kept its bow headed into the wind so that the stern well could provide some protection from the waves. Finally, all lines connecting the ship and the sub were cast off. *Deep Quest* was on its own.

As soon as we were clear of the ship, the waves increased and sea water sloshed over the deck of the sub. The experienced divers, knowing what was coming, had quickly moved to sit on the top edge of the conning tower.

But this was my first dive and I didn't know what to expect. So my reactions were a little slow, and my shoes and trouser legs were thoroughly soaked. The water itself was not uncomfortably cold, but my feet were wet and cold for hours during the dive. Next time I'll know enough to take along an extra pair of dry socks to avoid this discomfort.

The pilot stopped the thrusters when we were about 100 meters (approximately 100 yards) away from the ship. We sat dead in the water while further safety checks were made. With no way on—in other words, with no forward or backward movement—the sub wallowed and rolled with a sickening motion even though the sea was relatively calm, because of its rounded bottom and sides. Fortunately, no one got seasick, although I must admit that my stomach felt queasy for a little while.

Final preparations for the dive were made by two scuba divers, who rode out from the mother ship in an outboard-powered 18-foot Boston Whaler. They swam around and under *Deep Quest*, checking to see whether anything had been damaged during the launching. Then they removed lens caps from television and other cameras, took out the safety blocks from various places, and lowered the forward scientific instrument frame into operating position. When all was ready, the divers put finger to thumb in a circle, giving the pilot the traditional signal that everything was O.K.

The pilot opened the hatch and climbed down into the diving chamber, and we followed one by one. When the pilot pulled down and bolted the hatch, I felt a twinge of uneasiness. For the next few hours there would be no escape. The clanging of that hatch somehow seemed a little more ominous than the mere shutting of an airplane door.

The diving chamber, or pressure hull, of *Deep Quest* is bispherical. It is composed of two steel spheres, each 7 feet in diameter, connected by an opening that is 3 feet in diameter. The conning tower is not part of the pressure hull system; it is flooded when the sub is under water. The walls of the pressure spheres are made of a special high-strength steel alloy, and are almost one inch thick to withstand the tremendous pressures during a dive. The spherical shape of the pressure hulls is the

Marine geotechnical vane shear probe assembly (left center) on Deep Quest *ready to dive to sea floor. Mother ship* Transquest *in right background.*

Without accessory equipment on the bow, Deep Quest *looks like a hungry monster of the deep chasing a scuba diver.*

simplest and best shape for withstanding external hydrostatic pressure.

At our dive site the bottom was just over 4,000 feet deep. When we reached the bottom the hydrostatic pressure on the hull would be equal to the weight of the water directly above us, about 1,800 pounds per square inch of hull surface or about 130 tons per square foot. *Deep Quest* was designed to dive to a depth of 8,000 feet, where the pressure is about 3,500 pounds per square inch or 250 tons per square foot. She had gone down to that depth several times before our dive.

The pressure hull had been hydrostatically tested in a shore-based pressure tank to pressures equivalent to a depth of 8,500 feet. The depth at which the hull would collapse has been conservatively estimated at 13,000 feet. What those estimates mean is that the sub could probably dive to depths greater than 13,000 feet before the hull would be crushed. As a matter of comparison, World War II military submarines could safely dive to depths of only a few hundred feet. How far down a modern nuclear sub can dive is a closely guarded military secret, but it is likely to be only a fraction of the depths to which *Deep Quest* can dive.

When I reached the bottom of the ladder, I had my first look at the inside of *Deep Quest*. I had to stoop a little in order to keep from hitting my head on the many fixtures, electrical cables and pipes attached to the curved ceiling. The pilot and copilot, who is also the navigator, sit side by side in the forward chamber. In front of them is an instrument panel something like the one in a jet airliner, with a bewildering array of gauges, dials, indicator lights, and switches. Just above eye level, the pilot and navigator have television screens that can be independently connected to any of five outside television cameras, two of which the copilot can remotely control to aim in different directions. These are the principal means of visual contact with the outside.

Launched in June 1967, *Deep Quest* was one of the early deep-diving research submersibles, and it was built with only two fairly small direct-vision viewports, about 5 inches in diameter and several inches thick. One of these viewports

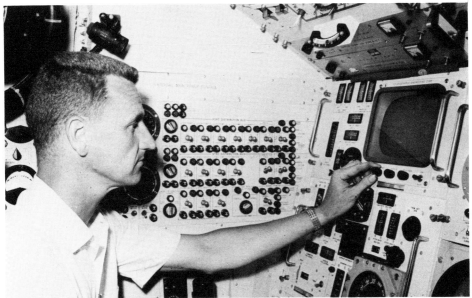

Pilot's side of forward sphere. Larry Shumaker adjusts the TV monitor controls.

Interior of Deep Quest, *showing instrument console, with pilot on left and navigator on right. Television monitor screens are seen in front of each man. Forward direct-vision viewport can barely be seen at center lower edge of photo.*

allows an observer to look directly down through the bottom of the rear chamber, but this viewport is seldom used because of the crowded conditions. The main viewport looks forward and downward from the front of the forward chamber. When the sub is near the bottom, the pilot can look down to the right of his feet through this viewport and can see the direction and speed of the sub's movement over the bottom. This can be important when there are scientific reasons for hovering in a stationary position against moderately strong water currents.

With the normal four-person complement, *Deep Quest* may be on the roomy side, as submersibles go. With me as the fifth person aboard, it was cramped. The forward sphere is pretty well filled by the pilot and navigator, their instrument console, television screens, and other bulky instruments. In the rear chamber, two passenger-scientists sit on small hinged seats that can fold up against the wall, wedged in between the racks of scientific instruments and a variety of pumps, tanks, and other apparatus that have to do with the operation of the sub.

During the several-hour dive I learned that there is no comfortable place for the fifth person. There were several possibilities and I tried them all. I could sit on the rim of the opening between the two spheres, but it was a narrow metal rim. I could sit on the floor, but there was little space for my legs and the steel flooring was cold—a few hundred feet down, the water temperature outside was near freezing. If I stood upright in the center of the rear sphere, my head was in the space around the access hatch. Unless I stooped in an awkward and tiring position, all I could see was steel just inches from my nose in all directions. Or when there was something to see outside, I could lie on the bare steel floor, crammed in at the front viewport.

It was not what you might call luxurious, lying face down on cold steel. The space was so narrow that I could not shift around, and it was so short that I had to keep my legs bent. In that posture my knees were jammed against a steel partition

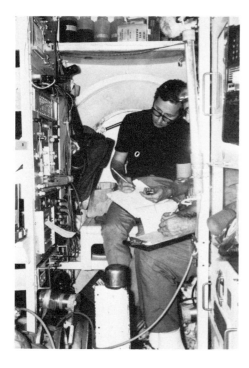

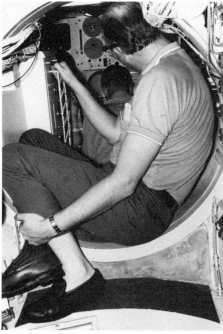

Rear sphere of Deep Quest *with scientific instrument racks, showing crowded conditions. Professor Terry Hirst is seated facing forward. Second man is almost out of view at right.*

An engineer installs and checks out the wiring for the electronic scientific equipment placed in the sub for this series of dives. The circular hatch beneath him gives access to the 360° optical periscope—the second viewport—which was not used on this dive because of crowded conditions.

The author curled up in the hatchway between the forward and rear spheres of Deep Quest. *As the fifth person in the sub, he has no regular place or seat.*

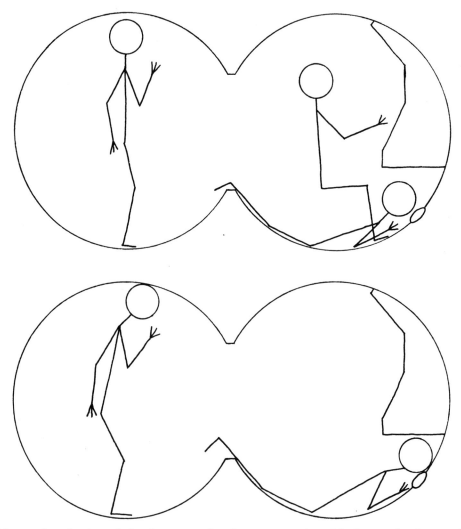

Upper sketch. An adult of average height can comfortably lie at the forward viewport or stand in the rear pressure sphere.

Lower sketch. A tall person has to stoop when standing and finds the viewport down around his chin.

and my shins extended over an unyielding metal edge. Remember, too, that this was the inside of a sphere I was lying on. If you want to get some idea of what it was like, lying on that floor that was curved the wrong way for comfort, try lying face down with your stomach touching the floor and the rest of you at assorted heights in the air. Makes you feel like the runners on an old-fashioned rocking chair.

In this position a person of average height would have no trouble seeing out of the small viewport. But I stand six feet four inches tall, and the viewport was down around my chin. I had to go through further contortions to see out at all, and then I could not look through my glasses but had to peer over the tops of the frames.

I found that at most I could stay in this position for maybe ten or fifteen minutes. Then my muscles cramped and I had to wriggle backward out of the slot between the pilot's and navigator's seats. As soon as I gave up the place, someone else eagerly moved in to take a turn at this small window.

I did change places with the other scientist-observers from time to time, but there was barely room to move around. If anyone wanted to shift his position, put on or take off a sweater, or even stretch an arm or a leg, everyone else (except the pilot and navigator) had to shift around to give the moving person room. You know what it's like when two people try to sleep in a bed that's only just wide enough for one. If either person turns over, the other one has to turn over too.

Did you ever see the mockups of the spacecraft vehicles? Remember how cramped they looked? *Deep Quest* actually has less free volume per man than the Mercury and Apollo capsules, and it has about the same volume per man as the Gemini capsules. Of course, *Deep Quest* is designed for a dive that lasts a day or less—a normal dive is 8 to 10 hours—whereas the spacecraft capsules are designed to be occupied for a number of days.

Deep Quest has about 45 cubic feet of free volume per person when four people are aboard—a space 3 feet by 3 feet by 5 feet. When five are aboard, each person has only 36 cubic feet. By contrast, a coffin has about 25 cubic feet and a federal prison cell has about 500 cubic feet per person. Anyone with a tendency toward claustrophobia—in other words, anyone who is afraid to be closed up in a small space—should never go down in a research submersible!

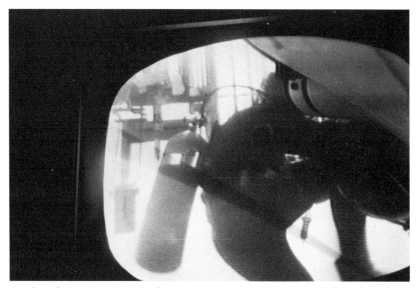

Scuba diver seen on television monitor screen performing last-minute safety checks prior to dive.

We continued to float at the surface for a few moments while the pilot and copilot checked out all systems and instruments. The pilot tested each thruster and swiveled the TV cameras so he could see them in action.

Deep Quest was designed to move under power in any direction—forward, backward, up, down, or sideways to left or right—as well as to turn sharply. Power thrusters were mounted in each direction to accomplish this. Each thruster consists of a 7½ horsepower electric motor and either a three-blade propeller or a water pump and jet nozzle. Two fixed axial thrusters produced forward motion or, when run in reverse, backward motion. Another two fixed thrusters provided up and down motion.

Just before the dive started, as a final step the pilot trimmed the submersible. Trimming consisted of shifting weight within the sub to balance it fore and aft as well as sideways. This was accomplished by pumping mercury from one tank to another within a system of small tanks and connecting tubes. Mercury is an ideal substance for a trim system. It is a liquid, so it can be easily moved from one position to another by pumping, and it is 13.55 times as heavy as water, so it takes only a small volume of mercury to shift considerable weight. And a double reaction is achieved because weight that is removed from one side is being added to the other side.

Because larger ships have far more space and can accommodate larger tanks, they do their balancing and trimming by pumping water instead of mercury from one tank to another. Also, mercury is expensive, and the large amounts needed for trimming a cargo ship would cost too much.

Finally, we were ready to go. The pilot trimmed the sub into a nose-down attitude by pumping additional mercury into the forward trim tank. This was the moment we had been waiting for.

Schematic cross-sections of Deep Quest, *showing dimensions and locations of major components. Transfer bell and man-in-sea module were never actually installed.*

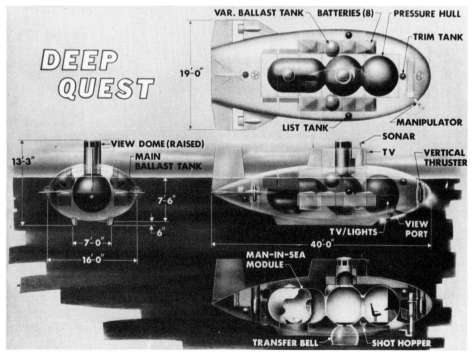

Descent to the Bottom

THE PILOT APPLIED FORWARD POWER TO THE TWO AXIAL THRUSTERS and we began to move. He decreased the sub's buoyancy—its tendency to float—by letting sea water into the main ballast tanks. This added to the overall weight of the sub, and we began to sink slowly. One way of descending to the bottom would be to let the craft just sink. If only a small amount of weight were added, the descent would be slow. If a lot of weight were added, the rate would be hard to control and it would be hard to stop the sinking when we were near the bottom. A better way is the active or flying mode of descent. Using the sub's stern planes as if they were the elevator flaps of an airplane, and also using the forward motion of the thrusters, the pilot guided the sub to the bottom as though he were flying an airplane. At any time he could adjust the stern planes and flatten out the descent or make it steeper.

We tilted forward and down at about 20 degrees from the horizontal—although it felt much steeper than that—and began to descend. For a few moments we could still see the scuba divers in the TV screens as they followed us for a few feet. Then they turned back to the surface and we were on our own.

The descent itself was relatively uneventful. There wasn't much to see after the first couple of hundred feet. At that depth the sunlight had faded away, and to conserve the batteries the pilot did not leave the exterior flood lights on all the time. Most of the life in the ocean is either near the surface in the sunlit portion, or at the bottom. In between there is not very much to be seen.

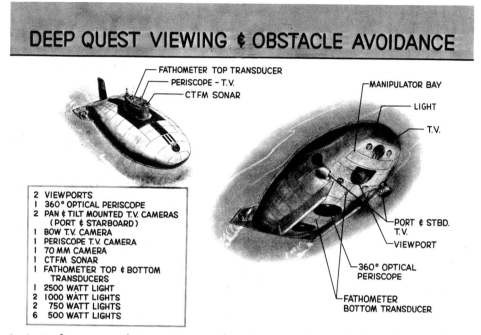

DEEP QUEST VIEWING & OBSTACLE AVOIDANCE

FATHOMETER TOP TRANSDUCER
PERISCOPE – T.V.
CTFM SONAR

MANIPULATOR BAY
LIGHT
T.V.

PORT & STBD. T.V.
VIEWPORT
360° OPTICAL PERISCOPE
FATHOMETER BOTTOM TRANSDUCER

2 VIEWPORTS
1 360° OPTICAL PERISCOPE
2 PAN & TILT MOUNTED T.V. CAMERAS
  (PORT & STARBOARD)
1 BOW T.V. CAMERA
1 PERISCOPE T.V. CAMERA
1 70 MM CAMERA
1 CTFM SONAR
1 FATHOMETER TOP & BOTTOM
  TRANSDUCERS
1 2500 WATT LIGHT
2 1000 WATT LIGHTS
2 750 WATT LIGHTS
6 500 WATT LIGHTS

Artist's drawings of Deep Quest, *showing viewing and obstacle avoidance systems.*

The steel walls of our pressure hull soon became cold to the touch. Water temperatures go down quickly with depth and are near the freezing point below several hundred feet. As the steel walls cooled, moisture condensed on them and there were a few annoying drips down the backs of our necks. When the first drop of water hit me, I wondered. Was the hatch tightly closed? Had we sprung a leak? Should I tell the pilot? As I looked around, I saw drops of water like dew all over the walls, and then I knew it was just condensation.

Soon we all wanted to put on sweaters, and there was much shifting and moving to accomplish this. In the cramped space we couldn't all put on sweaters at once, and we had to help each other and take turns before we were all properly dressed for the artificial climate inside *Deep Quest.*

Air pressure inside the sub is kept at normal sea level pressure, approximately 14.7 pounds per square inch, to provide what is known as a shirtsleeve environment, although we were in sweaters. A shirtsleeve environment really means that air pressures are kept near normal, so that there would be no need for special breathing apparatus or later decompression.

Deep Quest carries enough oxygen, in steel cylinders under high pressure, to sustain the normal crew of four people for 48 hours. As the air is consumed, oxygen is automatically added to maintain the normal 20 percent content that humans are accustomed to breathing, and the carbon dioxide gas that we exhale is absorbed from the air by chemical scrubbers. Fans circulate the air through the scrubbers and through electric heaters. Cooling would be needed in a submarine only in the rare ocean hot spots such as the one in the Red Sea where water temperatures get up to about 85 degrees Celsius (180 degrees Fahrenheit). There are also backup manual systems for controlling the atmosphere in case of power supply failure or equipment malfunction.

Oceanographic scientists may spend weeks or months at sea on surface ships. And when they do, chances are that many of them will grow beards—as a number of the *Transquest* crew had done. But you'll never see a beard or even a bushy mustache in a deep-diving submersible. The reason is simple. If the atmosphere in the sub were to become contaminated, the crew might have to put on full-face masks to breathe from an emergency closed-loop oxygen-demand system, and the masks cannot be sealed around a beard. One of the visiting scientists had to shave off a beautiful beard before he was allowed to make a dive on *Deep Quest*. There's not much chance that the emergency breathing system will have to be used. But it could happen. If the air did become unbreathable, would *you* want to take the time to shave before putting on the mask? And in an emergency the mask would be a life-saver because it would provide air for three hours, more than enough time to bring the submarine back to the surface.

One other aspect of life support systems is seldom mentioned. We were scheduled to be on the bottom for six hours, and it takes about an hour to descend and another hour to

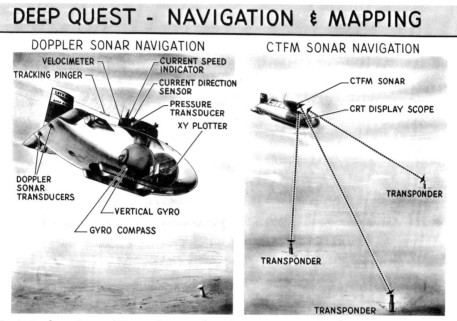

DEEP QUEST - NAVIGATION & MAPPING

DOPPLER SONAR NAVIGATION

- VELOCIMETER
- TRACKING PINGER
- CURRENT SPEED INDICATOR
- CURRENT DIRECTION SENSOR
- PRESSURE TRANSDUCER
- XY PLOTTER
- DOPPLER SONAR TRANSDUCERS
- VERTICAL GYRO
- GYRO COMPASS

CTFM SONAR NAVIGATION

- CTFM SONAR
- CRT DISPLAY SCOPE
- TRANSPONDER
- TRANSPONDER
- TRANSPONDER

Artist's drawings of Deep Quest, *showing navigation and mapping systems.*

ascend. Eight hours is a long time to suppress normal bodily functions. We had all been cautioned to use the head—or ship's bathroom—before boarding the sub, and we drank very little during the dive. But most of us did have to urinate, and the method was rather primitive: wide-mouth plastic bottles with tight-sealing caps. Larger submersibles—such as the *Ben Franklin*, which made an extended several-day submerged cruise drifting with the Gulf Stream—are equipped with more sophisticated toilet facilities.

Soon after we left the surface we were out of range for radio voice communication with the mother ship, as radio waves do not transmit well through water. The pilot switched over to a special underwater acoustic communications system. This operates in somewhat the same way as radio, but instead of a radio frequency carrier wave, it uses a high frequency sound that is inaudible to human ears. This carrier wave is amplitude modulated—its intensity or loudness is varied—by voice frequencies.

An acoustic transmitter puts about 100 watts of sound energy into the water by means of a transducer, a sort of underwater loudspeaker. That is more sound energy than most stereo or quadriphonic hi-fi's produce, but less than the amount put out by an amplifier for an outdoor rock band concert. An acoustic receiver, similar in function to a radio receiver, converts this modulated carrier wave into audible sound.

In this way, the pilot of *Deep Quest* can keep in touch with the dive master on the mother ship at the surface. The voice channel is not used all the time, as there is no real need for constant contact. Normally the sub's pilot reports in about every fifteen minutes. The rest of the time the channel is kept open in case of emergencies or changes in plan. For instance, if the weather at the surface should begin to produce high winds or waves that might make recovery at the end of the scheduled dive difficult, the dive master might tell the pilot to bring the dive to an end and return to the surface before conditions make a recovery impossible.

Another means that the sub and ship use to keep track of one another is sonar, a variation on radar. Sonar operates by injecting pulses—short beeps—of inaudible high frequency sound into the water, then detecting the sound pulses as they are reflected from objects in the water. The reflective targets show up as spots of light, called blips, on a circular screen. Sonar was developed by the navy for antisubmarine warfare, so that ships could pinpoint the location of submarines. The principles of sonar are now used by civilian and research ships for a number of search and navigational purposes.

The accuracy of sonar depends upon knowing very precisely the velocity (speed) of sound in sea water. Unlike radio waves, which travel at the same speed through air or through the vacuum of space, the velocity of sound varies with the substance it is traveling through, according to the density of that substance. In air, sound travels at a little over 1,000 feet per second. The speed varies slightly with factors that change the density of air, such as temperature, pressure, and moisture content.

Sound travels much faster in a denser medium such as sea water. At the surface under normal conditions, sound travels at approximately 4,915 feet or 1503 meters per second. Normal conditions are a temperature of 70° Fahrenheit (21° Celsius) and salinity (salt content of sea water) of 33.5 parts per thousand (33.5⁰/oo) or 3.35 percent (3.35%). The velocity of sound increases about 18 feet per second with each 1° Fahrenheit decrease in temperature, it increases about 30 feet per second with each 1,000 feet of greater depth (increased pressure), and it decreases about 16 feet per second with each part per thousand decrease in salinity.

For example, at a depth of 4,000 feet and a temperature of 35°F, in sea water of 35⁰/oo, the velocity of sound is approximately 4,853 feet (1,484 meters) per second.

Knowing the velocity of sound under the local conditions, sonar can indicate the distance to a target. It does this by measuring the time in thousandths of a second it takes for the sound beep to travel through the water to an object, be reflected off the object, and return to the sonar instrument. Beeps are sent out at regular short intervals by a rotating directional transducer. Targets or reflecting objects appear as blips or spots of light on the circular cathode-ray sonar scope, which is similar to a small television screen. The bearing of the target—its directional angle from the bow of the vessel— as well as its distance can be seen on the screen.

During the hour it took us to descend to the bottom, we prepared for our scientific mission on the bottom by checking out the operation of our special scientific equipment. In the preceding six months, project engineers had designed and built devices to measure the foundation properties of the sea floor; that is, the ability of the muds on the bottom to support heavy objects such as pipelines and sea floor mining equipment. These devices were instrumented probes—long metal rods—that were to be pushed into the bottom muds to depths of about 10 feet (3 meters).

When the probes were retracted they were housed in towers attached to the framework of the submersible outside the pressure spheres. These probes could be extended by rack and

DEEP QUEST WORK SYSTEMS

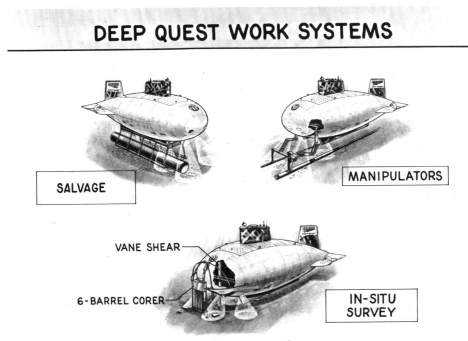

SALVAGE

MANIPULATORS

VANE SHEAR

6-BARREL CORER

IN-SITU SURVEY

Artist's drawings of Deep Quest, *showing work systems.*

pinion gears operated by a chain drive that was driven by electric motors. All this was controlled by electric cables leading to recording instruments inside the rear pressure chamber.

Everything had worked properly when tested in the laboratory. But since then the equipment had been flown air freight to San Diego and had been installed on the submersible. It had been impossible to check one vital aspect of the whole system's operation since it had left the laboratory. The probes could not be extended to their full length because they were under the sub, and the sub had been set down on a metal platform on the ship.

Now, when there was nothing but water beneath the sub, we extended the probes to their full length, testing all electrical circuits, motors, measuring devices, and recording instruments. Everything checked out perfectly on this dry run, and the probes were retracted.

As we descended, the surface sunlight had rapidly faded away. Below a depth of about 100 meters, the outside appeared almost pitch black. When we had completed the dry run instrument testing, I asked the pilot to turn off all the lights for a while. Only a dim glow of greenish light from the instrument panel illuminated the interior of the sub.

I stationed myself at the viewport. Now, as my eyes gradually adapted to the dark, I began to see scattered pinpoints of light. These were luminescent spots on small fish and other organisms. Living in total darkness, some fish have evolved the ability to produce light in special organs much the way lightning bugs glow in the dark.

This light, called bioluminescence, is used by some fish to attract other organisms for food. Anglerfish suspend a luminous lure in front of their mouths. Some deep sea organisms are attracted to these spots of light, whereupon the anglerfish captures and eats them. However, most midwater organisms have developed the opposite response and tend to swim away from light.

We turned on the outside floodlights in an attempt to see the fish with the luminous spots. But most of them swam away so fast that we hardly caught a glimpse of them and had no chance at all to photograph them. It took several minutes for our eyes to become adapted to the dark again.

Although the purpose of our dive was to study the bottom sediments, and observing the midwater life was not part of our mission, I had wanted to try to see the organisms that formed what has come to be called the deep scattering layer. When acoustic echo depth measuring devices came into general use after World War II, oceanographers were puzzled by false-bottom layers that appeared on some records. These midwater reflecting or scattering layers were strongest during the day, and appeared to break up or move upward at night.

The deep scattering layer (or false-bottom echo) is caused by the sound reflected by several types of organisms that contain gas bubble flotation devices. These organisms are seldom very numerous, but tend to be concentrated at certain depths.

The difference between a concentration of three or four organisms per cubic meter within the deep scattering layer, and only one or two organisms per cubic meter at other depths in the ocean, is sufficient. This diffuse layer of gas bubbles causes part of the sound (acoustic energy) to be reflected back to the surface, showing up as false bottoms on depth-finding records.

Schools of fish occasionally reflect enough sound energy to show up on sonarscopes or on acoustic depth records. But the deep scattering layer is too widespread and too consistently present to be caused by large fish. Attempts to net fish at depths corresponding to these layers have yielded only small numbers of very tiny fish, not enough to cause the reflection of sound.

Sound waves are reflected by materials having a density distinctly different from sea water, such as the mud and sediment at the bottom. These materials cause the speed of sound to change abruptly. Fish and other sea creatures are composed mostly of water. The speed of sound in water is almost the same as it is in the watery tissues of organisms, so that little sound is reflected back. However, most of the small fish netted at great depths exploded when brought to the surface. Further study showed that these fish contained swim bladders filled with gas. As the net was pulled to the surface, where the pressure was lower, the gas expanded and the swim bladder burst. It is the compressed gas bubbles within the swim bladder, different in density from the rest of the fish, that reflect the sound waves.

Each gas-filled swim bladder has a natural frequency depending on its size, much the way a small bell has a high-pitched tone or frequency and a large bell has a lower tone. When the sound waves striking the gas bubble are of that same frequency, the bubble resonates or vibrates and the reflection is much stronger than it would otherwise be. Echo sounders for measuring the depth to the bottom are seldom of the right frequency to resonate the objects in the deep scattering layer.

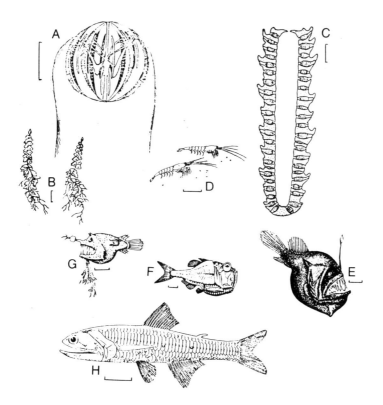

Some typical organisms found in deep scattering layers. Approximate relative size is indicated by a bar adjacent to each sketch that is one centimeter long in natural scale.

A. *Comb jelly (phylum Ctenophora)*, Pleurobrachia

B. *Pelagic siphonophore (phylum Coelenterata or Cnidaria)*, Nanomia, *with gas bubble float*

C. *Pelagic colonial tunicate (phylum Protochordata or Urochordata)*, Pyrosoma, *longitudinal section through colony*

D. *Shrimplike crustacean (phylum Arthropoda)*, Euphasia

E. *Deep-sea angler fish*, Melanocetus

F. *Hatchet fish*, Argyropelecus

G. *Deep-sea angler fish*, Linophryne

H. *Lantern fish*, Ceratoscopelus

Experiments with different frequencies of sound showed that 15,500 cycles per second (close to the tone of a high note on a violin) recorded the strongest reflections during the day at depths of 400 to 600 meters. The same frequency produced weaker reflections at night from much shallower depths. The strongest reflections at night were produced by much lower frequencies, but these lower frequencies hardly produced any reflections during the day.

The small fish that compose the deep scattering layer apparently migrate every day, up to shallow waters at night and back down to deeper waters in daylight. One explanation is that they hide during the daytime at depths just dark enough to make it hard for predators to see them. At night, under the protection of the dark, they migrate upward to feed on the more abundant shallow water life.

As our sub passed through the depths where the deep scattering layer should be found, we saw only a few widely scattered fish. In spite of the strength of the acoustic reflections, relatively small numbers of fish are involved. When a midwater trawl net with a 50-foot-diameter mouth is towed for an hour in a scattering layer, generally the catch amounts to less than ten pounds of marine life. A similar tow in commercial fishing grounds might yield thousands of pounds of fish.

After each of us had his turn at observing the luminescence in the deep scattering layer, the pilot again turned on the inside lights, and we went on preparing for our work on the bottom. To conserve battery power, the exterior floodlights were seldom used during the descent. If there were obstacles to be avoided, sonar would be of more use than visual observation.

After about an hour of steady spiraling descent, the pilot slowed our rate of descent and carefully watched the fathometer.

"Fifty feet off the bottom," he called out, as he switched on the exterior lights. We peered intently at the television monitor screens, and the navigator watched the viewport at his feet. Finally we began to see a shadowy something that must be the bottom. Some objects seemed to cast shadows from our bright lights. The pilot leveled off the sub, slowed the thrusters so as to barely keep us moving, and we settled slowly by our slight negative buoyancy.

On the Bottom

OUR EYE JUDGES SIZE AND DISTANCE BY MEASURING AGAINST familiar things in the surroundings. So the lack of familiar objects on the bottom, nothing to give a sense of scale, made it difficult to estimate distances visually. The bottom seemed a flat featureless plain, with nothing sharply defined. We still had some forward motion, and the scene below slid past us slowly. When we finally saw a starfish, it didn't really help because it could have a diameter of two inches or more than a foot. The echo fathometer was not much help either, as it does not work properly for such short ranges. And visibility was poor—we could see only a short distance ahead of the sub.

It is difficult to see very far under water even in the best of conditions. No sunlight penetrates this deep, and vision is completely dependent on artificial light. In even the clearest sea water, light from powerful floodlights does not penetrate far. The energy is absorbed by the relatively dense water.

Backscatter is an even more troublesome problem. Sea water is full of suspended particles that tend to reflect light. Some of these particles are so small that they take days or weeks to settle out of quiet water, and the slightest current or disturbance will stir them up again. Such particles include living microscopic single-celled organisms; dead organic fragments sinking from the abundant life in surface waters; and fine sediment of mud and clay, either stirred up from the bottom or sinking from the surface. Other larger multicellular swimming organisms also contribute significantly to backscatter.

Artist's view of Deep Quest *as it would appear when cruising near the bottom of the sea.*

The angle between the observer's line of sight and the illuminating light rays is also important. An analogy is with visibility during falling snow on land. In the daytime light coming from above makes a large angle with the line of sight, which is horizontal. As a result, visibility is reduced only slightly by the falling flakes. At night, however, the artificial lighting from the headlights of a car is nearly parallel with the line of sight. Most of the light is therefore reflected or scattered by the snowflakes, and visibility is reduced greatly even though the snow conditions are the same as in daytime. When headlight beams are high, visibility may be almost zero. But there is a noticeable improvement on low beams because the angle between illumination and sight is a few degrees larger.

On a submarine as with an automobile, it is all but impossible to place lights so as to make an appreciable angle to the line of sight, so backscatter always reduces visibility. To further aggravate the problem, the lowest 8 or 10 feet of water nearest the bottom is usually the most turbid (suspended material is most abundant), so right where we most want to see, visibility is the poorest. At our dive site, we guessed that we could see perhaps 25 to 30 feet. We had to rely on search sonar to tell where objects were at greater distances.

With exterior floodlights on, we descended the last few feet of our approach to the bottom under visual guidance. The pilot pointed the TV cameras downward, watched the TV monitor screens, and occasionally glanced down through the visual viewport near his right foot. Finally we touched bottom with a barely perceptible bump. We all watched the TV screens as clouds of sediment, looking for all the world like dust, swirled about us, stirred up from the bottom. For a few moments we could see absolutely nothing.

The "dust" we had stirred up is actually minute particles of mineral and organic matter. They are less than sixteen thousandths of a millimeter (0.016 mm), or about six ten-thousandths of an inch (0.0006 inch) in diameter. After a few minutes, the bottom currents carried away the cloudiness and we could see again.

After that the pilot was careful never to touch bottom with the submarine until our work required it. Instead he maneuvered to hover like a helicopter a few feet above the bottom, carefully staying above the backwash effects of the downward directed vertical thrusters. To make hovering easier, he adjusted the buoyancy of the sub to where we were barely negatively buoyant; that is, so that we tended to sink toward the bottom very slowly, perhaps an inch per second.

For the descent we had been strongly negatively buoyant. To decrease that our pilot used compressed air to force a little sea water out of the main ballast tanks. Now it took very little power—the thruster propellers turned very slowly—to hover above the bottom. We could stay over the exact spot on the bottom that we wanted to observe.

Although gentle bottom currents, at most a few inches per

Test site markers, one with attached float, seen on TV monitor.

second, tended to drift us away, only a little power applied to the forward thrusters was enough to maintain our position. If there was a current, we usually headed the sub into or away from the current, but we could also move sideways with the lateral water jet thrusters. Water jets are simply pumps that force a stream of water out of a pipe that is oriented crossways to the sub. The reaction of this stream against the outside water moves the sub sideways. Using a combination of right or left sideways thrust and forward or backward thrust, the pilot could maneuver the sub in any direction.

The pilot reported to the surface ship that we were on the bottom and ready to commence our research tasks. Our first task was to find a pair of large blocks of concrete that had been placed on the bottom a few months earlier. These were to serve as reference points for bottom navigation during our movements along traverse lines. To help us find the concrete blocks, acoustic devices called pingers were anchored near them. A pinger is a battery-operated device that emits pulses or beeps of sound that can be "heard" by the sub's sonar apparatus.

On the circular sonar screen the center represents the sub, and straight ahead is at the top of the screen. When we are close enough to hear the pinger, its signal shows on the screen as a bright flashing line running from the edge of the screen toward the center. The position of this line on the edge of the screen indicates the relative direction from which the sound is coming. The pilot has only to turn the sub to port or starboard until the pinger signal is lined up with the top of the screen and the sub is then pointed toward the pinger.

In normal operation, sonar can provide both bearing and distance to an object. But in this situation of homing in on a pinger, its distance from the sub cannot be determined. In normal operation, the ping is emitted from a source on the sub: the sound pulse travels from the sub to the object and bounces back, and the distance is calculated from the time it takes for this two-way journey. But, when the pinger is on the object, we have no way to measure the time it takes for the ping to reach the sub simply because we don't know when it started.

Our sonar picked up the pinger's signal right away, and the pilot had no difficulty in guiding the sub to the concrete block markers. The signal from the pinger is strong enough to be detectable at a distance of at least two miles, far greater than the range at which our search sonar could pick up reflections off the concrete markers themselves. Had we started our search beyond the audible range of the pinger, finding the concrete markers would have been vastly more difficult.

Pinpointing an exact spot on the open ocean is not easy for a surface ship. Yet it was important that the mother ship *Transquest* launch the sub near the right place. Ordinary navigation—getting from one port to another—does not require great accuracy in locating the position of a ship on the open ocean. Traditional astronomical methods are inadequate for oceanography. They must be done at night, when the stars are visible, or exactly at noon local time, whereas we needed our exact location at 0900 hours. Radio methods did not serve our purposes either. Methods such as loran or decca are barely good enough under ideal conditions near shore. Clearly, large parts of the open oceans cannot be covered by these methods.

Our dive site had been selected close enough to shore so that tall shoreline landmarks such as lighthouses and watertowers were in sight, and the skipper of *Transquest* could take visual compass bearings on them. Visual bearings were taken at intervals all day long to enable the *Transquest* to remain in the vicinity of the dive site. This was necessary to stay within voice communication range of the sub, as well as to help in some of the sub's bottom navigation procedures.

Down at the bottom we homed in on the pinger, and within ten minutes the two reflection blips of the concrete markers appeared on the edge of the sonar screen. We moved slowly along the bottom at the sub's cruising speed of two knots or just under three miles per hour, about normal walking speed. At last we came close enough to see the concrete blocks on the TV monitor screen.

Part of our mission on this dive was to survey a rectangular area a few hundred meters on each side, to establish the locations for probing the sea floor with our instruments on this and later dives. We wanted a uniform flat bottom on which to begin our work, and as we moved around the margins of our rectangle and later crisscrossed it in a series of traverses, we would carefully observe the bottom to make sure there were no unexpected and unusual features that might interfere with our plans. At the same time we would make a detailed map of the area on which to plot the exact positions of all our probe tests. This would require extraordinarily precise navigation, and we planned to use a method of relative positioning.

Our locations on the bottom would be determined in relation to fixed reference marks. Our primary sea floor references were the concrete blocks and their pingers. To complete a reference baseline, we needed another marker at a known distance and direction from the concrete blocks. A special pinger was to be dropped overboard from the surface ship. An attached weight would cause it to sink rapidly to the bottom. Then we in the sub were to locate it and measure its distance and compass bearing from the concrete blocks.

This special pinger, called a transponder, pings only after being triggered to respond by receiving an acoustic signal from the regularly spaced pings of the search sonar. With this arrangement, the signal from the transponder is automatically detected and shown as a blip on the sonar screen, giving both bearing (angle from sub's bow) and range (distance from sub to transponder).

When we were ready on the bottom, the pilot called the surface ship on the voice transmitter to inform the dive master that we were on station. Soon the message came back from the ship that the transponder had been dropped overboard. The transponder, a cylinder about 4 inches in diameter and about 2 feet long, had attached weights that would carry it to the bottom in about 15 minutes. The transponder also carried a float with enough buoyancy to lift it back to the surface when the weights were not attached.

We waited the 15 minutes. Then the copilot pressed a switch sending a coded series of beeps to turn on the transponder. Without realizing what I was doing I held my breath as I watched over the copilot's shoulder for the answering flash on the sonar screen. It should have taken only a few seconds to respond, as the *Transquest* could not be more than a few hundred meters away from a position directly above us, and there were no strong currents to keep the transponder from falling straight down. But I couldn't hold my breath long enough—there was no answering flash from the transponder.

After repeated interrogations without response, we were forced to conclude that the transponder had failed to operate. One of the frustrations of oceanographic research is that no matter how carefully and thoroughly instruments are checked and tested before going into the ocean, malfunction is common. The ocean environment is harsh, and the extremes of high pressure, low temperature, and corrosive electrically conducting sea water are hard on electronic instruments in particular. This state of affairs was just another application of Murphy's law: If anything can go wrong, it will.

Without the transponder, we could not perform our tasks with the precision of navigation needed to serve as a basis for the research to be done on later dives. The transponder costs about $4,000. We did not have a spare and the project could not really afford the loss. On the other hand, searching for the transponder would be like searching for a needle in a haystack. How much time could we afford to devote to such a search? A dive on *Deep Quest* costs more than a thousand dollars per hour when all operating costs and salaries for the crews of the sub and mother ship are added up. The transponder was a fairly small object that could be anywhere within a radius of about a mile.

We talked about it for a few minutes on the sub. The pilot has the final say on anything that affects the safety and operation of the sub and crew, but the scientific party must recommend any necessary changes in plan with due regard to how it will affect the research program. We agreed that we shouldn't abandon the transponder without making an attempt to find it, but that two hours was long enough to spend on the search. The pilot explained our decision and plan to the dive master on *Transquest*, who gave his permission to proceed.

It was soon evident to me how difficult a task it was to make a systematic search of even a small area of the sea floor. As we couldn't see very far, we would have to rely on the search sonar to reveal the presence of objects on the bottom. Whenever we saw a strong blip on the sonar screen, we would move closer to look at that target. If it turned out to be something other than the transponder, we would continue the search.

To keep from wasting time by looking at the same object more than once, the copilot plotted the track of the sub continuously on a chart, and marked an X through each spot on the chart where we had looked at a sonar target.

The pilot guided the sub on a search at increasing distances from the concrete block markers. One logical pattern would be a spiral of ever widening circles, but following a curving course and keeping a specified distance from a marker is difficult. It is much easier to steer a straight line by watching a

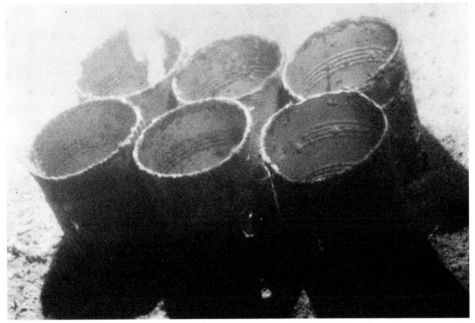

Typical of the larger trash on the bottom is this rack of 6 buckets or paint cans, each about 30 centimeters in diameter.

compass. So we began to trace a series of squares, moving a little farther away from the concrete marker when we had completed each square.

Our track lines were about 100 meters apart. This gave a little overlap for the distance at which our sonar could register the presence of an object as small as the transponder, even if it were lying on its side on the bottom instead of being oriented vertically as it should be.

We had to look at every small target blip that showed up on the sonar screen, so those of us in the scientific party had an unusual opportunity to observe an area of sea bottom intensively. Although most of the target inspection could be done more quickly with the television system, we took turns standing watch at the forward visual port.

Any object with a hard surface that reflected the sonar sound pulses could be a sonar target. Most of them turned out

Large piece of junk on the bottom, probably a rusted and encrusted barrel.

to be things like gallon paint cans, buckets, a small steamer trunk, a locker or metal cabinet, and many, many beer cans. We saw other objects that were not sonar targets, such as a shoe, several feet of copper wire, a large crumpled sheet of plastic, a number of flat, square, rusty, unidentifiable objects, and an assortment of unrecognizable objects.

To those of us who are engaged in environmental research, it was disconcerting to see this much junk on the ocean floor 20 miles away from the coast in 4,000 feet of water. Although the area is within easy reach of San Diego in a power boat, it is not near or on the way to a sport fishing area or a dumping ground.

Not all junk on the bottom is detrimental to fish and other marine life. Some kinds of junk provide shelter from predators or growing surfaces for many forms of life. But there are probably limits to the amount of junk that a given area of ocean floor can safely absorb. Scientists have speculated that we may learn what those limits are only after we have exceeded them, and by that time at least some of the effects may be irreversible.

One of the problems with junk on the deep sea floor is that it seems to last forever. In most places on land or in shallow water, junk either rots, rusts, or decays. Furthermore, it is usually buried in a relatively short time by natural processes. Such waste materials are recycled naturally. However, the ocean depths are a natural deep freeze. Below a few hundred feet, water temperatures are barely above freezing in many parts of the ocean. Also, most of the oxygen has been used up from deep waters, and it is not replenished very rapidly. Both low temperatures and a limited supply of oxygen slow down bacterial decay, oxidation, and other natural processes of destruction.

Cold, deep, low-oxygen ocean water is a good preservative. On another dive on a mission for the navy, *Deep Quest* found and helped recover a World War II Helldiver torpedo bomber. After nearly 30 years in the deep ocean, the aircraft appeared to be in surprisingly good condition. However, as soon as it was exposed to air and began to dry out, it deteriorated, corroded rapidly, and soon fell apart.

Another example of the slow pace of decay in cold deep ocean water occurred with the sinking and recovery of the submersible *Alvin*. The pilot's lunch—a sandwich and an apple—were found intact and in so good a condition that it seemed almost good enough to eat, except for being soaked in salt water, after being in the sea for nearly a year. Thus the deep ocean bottom is practically a cold storage warehouse. In fact, seawater can be cooled to below the freezing temperature of fresh water and remain liquid because the high salt content acts much like an antifreeze.

Ocean-going ships commonly throw overboard or "deep six" all garbage and junk, and untreated toilet and engine wastes are pumped directly into the sea. The oceans are limited; they are not infinite sinks for waste disposal. The practice of indiscriminate dumping from ships is being curtailed, but has not yet been entirely stopped. The wide publicity given to the dumping of obsolete poison gas and radioactive wastes has discouraged that practice. However, many industrial plants are now applying for permits to dump spent acids and other toxic wastes at sea, as disposal on land becomes more restricted.

Some people think we worry too much about poisoning a few fish or a few birds. What really worries the concerned scientists are the possibilities of starting something that we can't stop. There is much we don't know about the irreversible consequences of seemingly small events and their effects on the human race. One such possibility is that by tampering with the oceans we might reduce the oxygen content of the atmosphere.

At some as yet unknown level of concentration of toxic substances in the ocean, the number of phytoplankton could be reduced by 10 or 20 percent. This can be done in the laboratory under controlled conditions, but in the vast expanse of the ocean we may not be able to detect it until too late. Phytoplankton, the floating plants of the sea, are mostly microscopic or small in size.

All the oxygen in the atmosphere is produced by the photosynthesis process in plants. Most of the world's plants live in the ocean, so killing off a fraction of the phytoplankton could reduce oxygen production by a few percent.

There is a balance between the amount of oxygen dissolved in seawater and the amount of oxygen in the atmosphere, but the rate of exchange is very slow. By the time a decline in atmospheric oxygen were noticed, it would be too late to do anything about the lost phytoplankton. The long-term effects on humans of a relatively small decrease in atmospheric oxygen are totally unknown.

This is not meant to prophesy doom. It does point out the need for increasing our understanding of the many complex ecological interactions within and between the oceans and the atmosphere. We must learn to recognize the long-term consequences of our actions before we carry them too far. One seemingly small action may trigger a series of events with damaging effects that cannot be reversed. Although the ocean has a tremendous buffering capacity, or ability to neutralize acids and many toxic substances, no one knows what the limits might be for absorbing without detriment the many new waste chemicals now being dumped into the world's waters.

Our search for the transponder was frustrating. Time after time we interrupted our regular search pattern to have a close look at a sonar target, and always what we found was a piece of junk. It was beginning to get monotonous, and a square mile of ocean floor seemed awfully big.

Our time was running out. Then, just before the two hours were up, we found the transponder! Our systematic search paid off—but there was a lot of luck involved too. Just another interruption, another target to be checked. We moved closer. At 15 meters (50 feet) we could see it. The transponder was standing upright in the water, hanging down from its spherical floats, and attached to its anchor lying on the bottom. And it still didn't work—wouldn't respond to the triggering signal from the sub.

Larry, the pilot, usually very matter-of-fact, sounded almost excited when he reported our discovery to the dive master on

The unresponsive transponder found at last.

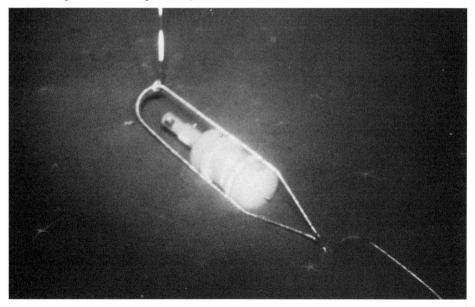

Pinger (vertical cylinder in lower center) with attached floats. Two floats are hanging down, apparently filled with water. Mechanical arm of sub is seen at right.

the *Transquest.* We were instructed to send the malfunctioning transponder back to the surface.

The method for doing this was rather ingenious. The pilot maneuvered the bow of the sub right up to the transponder. He then took a flat metal control box off a wall bracket and placed it in his lap. A thick multiwire cable led to a wall junction box. The top of the control box bristled with toggle switches and two larger joy stick controls. The pilot flicked a couple of switches, grasped a joy stick lever with each hand, and moved them slowly.

The copilot adjusted a television camera to that we could see the transponder clearly in the TV monitor screen. Slowly a mechanical arm moved into view from the side and reached down below the transponder. The hand on the mechanical arm touched the cable from the anchor weight and moved slightly. Suddenly the transponder moved straight up and out of sight.

Using remote controls from inside the sub, the pilot had operated a manipulator arm with a cable-cutter hand, with which he had cut the transponder free of its anchor weight. The attached float carried it back to the surface in about 25 minutes.

Much of the useful work done at the bottom of the sea from submersibles is performed by mechanical manipulator arms. *Deep Quest* has two manipulators as part of its regular equipment. Each had several elbow and wristlike joints, allowing almost as much maneuverability as a human arm. The hands are interchangeable tools, including clamps or pincers, cable cutters, wrenches, drills, hacksaws, and welding torches. Operating the arms from the multiswitch control box is very tricky and requires a lot of practice. The various parts of the arms can be moved in five different ways (five degrees of freedom) compared to the seven different movements of the human arm. Each movement can be made along one axis at a time for a jerky mechanical motion, or several movements can be done simultaneously for smooth flowing motion.

Our sub's mechanical arms are powered by a hydraulic system. Hydraulic means liquids in motion. An electric motor drives a high-pressure oil pump that circulates oil under pressure through flexible hoses and copper tubes to hydraulic motors that move the arm joints and tools. A hydraulic motor is simply an oil pump in reverse, converting moving liquid to mechanical motion. Hydraulic power systems are mechanically simple and adaptable, and they can produce either rotary or linear motion at variable speeds without gears. When fully extended to a length of 6 feet, *Deep Quest's* arms can each lift 100 pounds.

While we waited for the workmen on *Transquest* to retrieve and fix the transponder, we in the sub ate lunch—sandwiches and hot coffee from a thermos. The success of the search, the warmth of the coffee, and the nourishment of food did much to lift our mood.

At the surface, the transponder was spotted soon after it bobbed up. Scuba divers went out in the powerboat, recovered

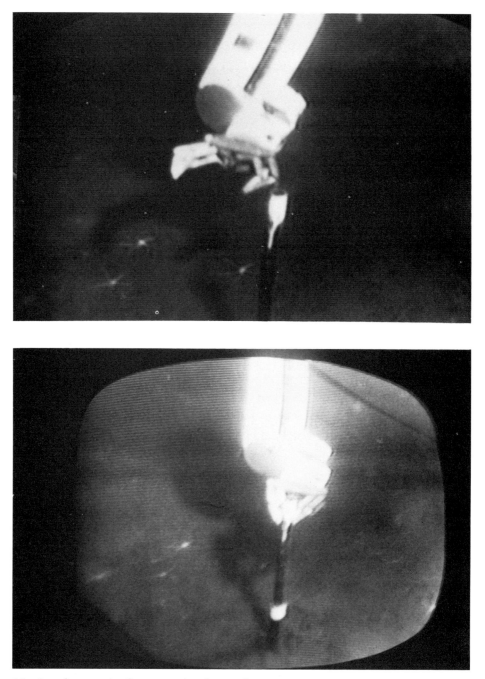

Moving the manipulator arm and jaws by remote control from "near" in upper photo to "grasping the object" in lower photo is a tricky maneuver.

With its two mechanical arms extended forward, Deep Quest *appears to be a gigantic bug.*

it, and brought it back to the ship. Most frequently when electronic gear fails to operate in the ocean, a seal that was supposedly watertight has allowed seawater to enter at the great pressures, and the seawater, which conducts electricity, has caused short-circuit in the equipment. Engineers on *Transquest* opened the transponder container, but found no water or dampness inside, no evident reason for the malfunction. They reassembled the unit, tested it with an interrogator pinger, and found that it responded properly. They attached another anchor weight and dropped it over the side to sink down to us again. The dive master told the sub pilot it was on the way down.

When we thought it had had time to reach the bottom, we tried again to interrogate it. And again there was no response. We could only conclude that although it didn't leak under pressure, the surface test was inadequate. Something was not operating properly—perhaps because of the low temperature or perhaps because of the high pressure. There is much room for improvement in the design of oceanographic equipment, and much to challenge the innovative engineer.

This second failure of the transponder was discouraging. We quickly decided that making another search was not the best way to use our time on this dive, and we left that task for a later dive. There was one other task that should be performed on this dive: replacing the batteries in one of the reference pingers. The two pingers had been planted three months ago, along with the concrete block markers. The pingers had a battery life of six months. We were to release one pinger and let its floats carry it to the surface. There the batteries could be replaced, and on the next dive the pinger could be re-planted. If this were not done, the batteries might give out before the next series of dives at this location, scheduled in four months. Without a working pinger the concrete block markers might be difficult to find again.

We homed in on the nearest of the two pingers. Two of its five floats were hanging down. They had apparently leaked and were no longer buoyant. It took several minutes to ma-neuver around and cut the waterlogged floats free with the cable cutters and the manipulator arm. The other three floats must also have leaked, as they did not have enough buoyancy to lift the pinger taut against its anchor line. If it were cut free from its anchor, it would not float to the surface.

The pilot called the dive master, explained this problem, and decided that the best way to recover this pinger was to carry it back to the surface. However, as the special basket for this purpose was not attached to the sub on this dive, he would postpone the recovery to a later dive in this series.

We still had nearly three hours that we could spend on the bottom, and we asked ourselves what we could do in that time that would be of most use to later dives. It didn't take us long to decide that we ought to have a close look at the area, even if we couldn't navigate the traverses with the precision that the transponder would have given us. We would proceed 1,000 yards due north, then due east for 600 yards. The copilot would calculate the distance we had gone by dead reckoning. In other words, he would work out the distance without checking the sun or the stars. Instead, he would multiply our speed by the elapsed time, and would crosscheck with sonar fixes on the concrete blocks.

One of the purposes of making these traverses was to obtain a photographic record of the visual bottom characteristics. We would take a bottom photograph every 30 seconds. There were two exterior cameras on *Deep Quest:* a 35 mm for color slides and a 70 mm for black and white photos. Both were housed in watertight pressure-proof containers, were battery-operated for film advance and shutter release, were fixed in position and focus to cover an area that could be seen from the forward viewport, and were synchronized with an exterior strobe light. We had also rigged another 35 mm camera with black and white film inside the pressure hull to photograph the TV monitor screen.

We took turns watching at the forward viewport. The copilot called out *mark* every 30 seconds and wrote the frame number on the chart of our traverse. The observer at the port-hole pressed the cable release every time he heard the call. This systematic approach did not last long, as the most interesting objects to photograph almost always showed up between the prescribed intervals. Soon the observer was calling out *mark* every time he photographed another fish or peculiar form of life, and the recorded sequence never quite matched the number of frames actually shot. But we did get some interesting photographs.

The others in the sub had been on earlier dives and had seen all this many times before. The pilot had been on 110 of the

120 dives made so far by *Deep Quest*. I had never been deeper than about 60 feet in scuba gear, and my first views of the relatively deep ocean bottom were fascinating and magnificent.

The bottom sediment was much lighter in color than the darkish slate-gray I had expected, but I was told that the intense light from our lamps made the bottom appear lighter than it really was, almost a creamy white. On properly exposed color photographs, the bottom appears close to its true color, a medium brownish-gray. This was confirmed later when we looked at sediment samples in the laboratory.

The variety and abundance of life also impressed me. Somehow I had come to believe that the deep ocean floor was all but a biological desert, almost devoid of life. Perhaps this impression came from grab samples brought up from the bottom by surface oceanographic ships. These grab samples usually covered only about a square foot of bottom and seldom brought up anything live. Much of the bottom life is actively mobile and can of course move away and avoid a foreign object like a bottom sampling device. Early bottom photographs taken by automatic cameras lowered on cables were often poorly exposed and badly focused, and they did little to reveal the richness of bottom life.

In our dive area the most obvious animal life are starfish. These are mainly brittle stars, belonging to the class Ophiuroidea of the phylum Echinodermata. There appear to be three or four different types of brittle stars of varying size and color, some white, some pink, some gray. These may be merely growth stages of the same species, as it is somewhat unusual for several closely related and similar species of the same invertebrate animal to live in the same habitat. However, if they aren't in direct competition—if they eat different kinds of food, for instance—they may occupy different ecological niches while living in similar or overlapping territories.

The brittle stars were as dense as five or six per square meter and ranged in size up to a maximum of about 30 centimeters in diameter. Two or three other kinds of true starfish

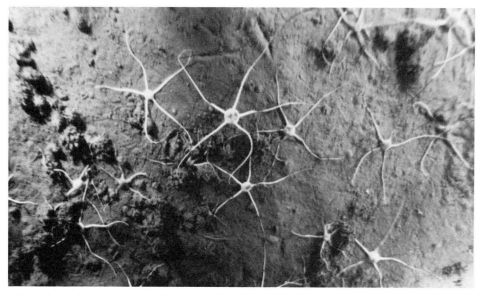

Abundant brittle stars (Ophiuroidea) on the bottom mud. For scale, the largest star in the center of the photograph is approximately 25 centimeters from tip to tip of its outstretched arms.

(class Asteroidea of the same phylum Echinodermata) were also present but much less abundant.

Other forms of life were much less common. We saw a variety of crustaceans, swimming or crawling creatures that looked much like shrimp, crabs, or lobsters (class Crustacea, phylum Arthropoda). Some of these appeared to have dug burrows in the bottom mud, as we could see low mounds and holes similar to those made by burrowing crustaceans in shallow water.

In some areas the bottom had a gentle hummocky appearance from the numerous burrows and piles of castings. Some of the burrows may have been made by worms, although they were seldom visible. We did see many tube worms. They looked for all the world like small feather dusters waving in the water until disturbed by the motion of the sub, when they retracted into what looked like plastic tubes buried in the mud.

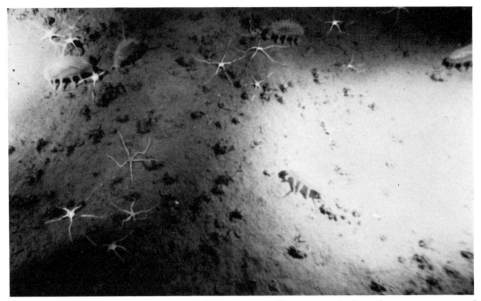

Several many-legged sea cucumbers, Scotoplanes, *with brittle stars on mud bottom marked by small mounds produced by unknown burrowing organisms.*

Closer view of Scotoplanes *being tumbled by the currents produced by the sub's thrusters.*

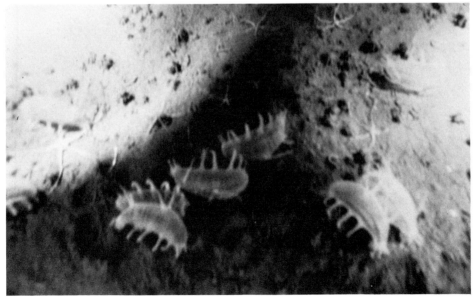

The moving sub caused gentle water currents that showed up as rippling movement in the organisms not attached to the bottom. The most obvious of these looked like small tumbleweeds rolling on the bottom. On closer inspection we saw that these were sea cucumbers, almost transparent bluish or pinkish sausage-shaped bodies a few inches long, covered with short thick tentacles protruding all over. These belong to the class Holothuroidea of the phylum Echinodermata, and are unique in that phylum for having little if any skeleton.

Other life forms were suspended in the water. Most were too small to be clearly seen, ranging in size from microscopic to a few millimeters. These were mere blurs of motion that contributed to the poor visibility through the water. A few of these planktonic (floating) forms were large enough to be seen—such as the almost entirely transparent, pulsating, walnut-sized jellyfish or comb jellies, more properly called ctenophores. These are so different from anything else in the ocean that biologists classify them in a separate phylum, Ctenophora.

Occasionally we saw fish of various shapes and sizes. Many seemed to avoid the bright lights from the sub, and we only caught glimpses of these as they moved swiftly away. Most of the fish appeared to be indifferent to the light, and a few were attracted to it. Probably most were entirely blind, living in eternal darkness.

It was very difficult to estimate the size of the fish we saw, as generally nothing familiar was visible in the background to provide a scale. Some may have been a meter or more in length, but most ranged from a few centimeters to several centimeters in size.

There were no plants at all. Plants cannot live at this depth because of the total absence of sunlight, which all true plants need for photosynthesis. Plants are usually the first link in a complex food chain, as they alone can make food from inorganic materials. For the animals living on the deep sea floor, the first link in the food chain is probably the dead organic matter sinking down from the surface waters. Most deep

A bright pink shrimplike crustacean.   *A large asteroid sea star.*

A long eellike hag fish.

ocean dwellers are thus scavengers. Some forms are carnivores that feed on other living animals. There are no true herbivores or plant-eating animals in this environment.

Some deep sea animals resemble plants in having stalks, branches, and what appear to be flowers. Most of these are close relatives of corals such as sea-whips, sea fans, and sea pens (class Anthozoa of the phylum Coelenterata). Smaller plantlike organisms, tufted mosslike colonies of minute animals, are the so-called moss animals belonging to the phylum Bryozoa.

Many of the deep sea forms of life have not yet been adequately studied by marine biologists. Most have not even been classified or given scientific names. The bulk of the marine biologists now graduating from colleges and universities work with shallow water organisms. One reason is that most of the commercially important fish, as well as the vari-

A large true bony fish (Osteichthys) of unknown genus and species, perhaps 100 centimeters long.

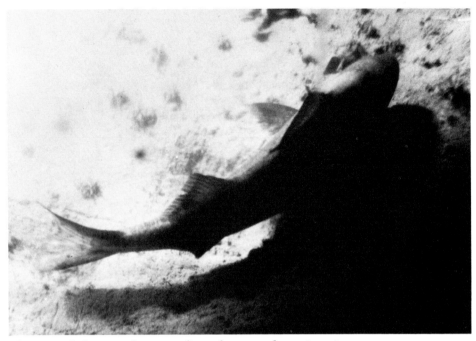

The same fish as in the preceding photograph, swimming away.

ous elements in their food chains, live in the upper few hundred feet of the ocean. And despite an intense interest in marine biology of the deeper oceans, biologists have been frustrated in obtaining good specimens of deep sea animals.

While we could see and even photograph many organisms from the submersible, we couldn't capture specimens. One of the most frustrating aspects of observing marine life from a submarine is that you can see but you can't touch. When scuba diving in shallow water you can use your hands to pick up objects and turn them over to examine them closely. You can use a net to capture swimming organisms, and you can put specimens in a bag to be carried back to the surface. But in a submarine there's not much you can do but look.

Most research submarines do have mechanical arms to which various tools can be attached, and some useful work can be done. However, the mechanical arms can seldom be moved swiftly enough to capture moving organisms. As a further complication, few specimens can be brought back alive from very deep water because the change in pressure is too much for them. Many organisms that are used to the great pressure at the bottom of the sea will actually explode as gases within their bodies expand when the outside pressure is decreased.

Some research submarines are designed for work at relatively shallow depths, to a maximum of about one thousand feet, and can carry scuba divers. Such a submarine has two separate chambers: one for the pilot and crew and one for the divers. The atmosphere in the pilot's chamber is kept at normal sea-level pressure, but in the divers' chamber the pressure can be increased to equal or slightly exceed the pressure outside the sub. Then the divers can open a special hatch in the floor of their chamber and leave the sub. The increased air pressure in their chamber keeps the seawater from coming in. The divers can therefore perform their tasks in the water away from the sub, and can then return and re-enter the sub.

After they close their hatch, the water can be pumped out of their chamber and the divers can take off their scuba breathing apparatus. However, the air pressure in their chamber must be kept at the high level or they would be likely to get "the bends," the painful and sometimes fatal condition produced when a diver breathing high pressure air moves into a lower pressure environment too rapidly. Gas in solution in his bloodstream forms bubbles that collect in the joints or in the capillaries of the brain and cause great pain and (possibly) death. The pain in the joints leads to a "bent over" posture, which gives this condition its name.

The divers must decompress gradually over a period of time by slowly lowering the air pressure in their chamber on the sub or in a decompression chamber back aboard ship before they can safely breathe at normal pressures again. Because the two chambers are at greatly different pressures, the divers cannot enter the pilot's chamber or vice versa. A submarine of this type is said to have diver lockout capability.

Our submarine *Deep Quest* does not have diver lockout

A large fish swimming near one of the site markers, a concrete block 100 centimeters on each side.

The manipulator arm, upper left, moves in to pick up an old brass ship's lantern found on the bottom.

capability, although that feature was part of the original design plan. *Deep Quest's* dive missions are usually to depths greater than a free diver can go. Even with special gas mixtures (helium and oxygen, for example), scuba divers can descend safely to depths of only about 1,000 feet.

Under carefully controlled conditions in a laboratory pressure chamber, men have successfully been subjected to pressures equivalent to depths of slightly greater than 1,500 feet. It is unlikely that divers will actually go that deep. The amount of time they can spend on the bottom doing useful work is so small compared to the time required for decompression that such dives are uneconomical if there is any other way to accomplish the task. The obvious other way is in a submarine equipped with versatile mechanical arms and a set of attachable tools.

Considerable research and experimentation are now being done to devise methods and equipment for capturing and

World War II Navy Hellcat dive bomber, found off California coast by Deep Quest. *The airplane was lifted to the surface by a ship's cable that was attached by the submersible. Scuba divers follow the plane the last few feet to the surface.*

bringing to the surface in good condition the multitude of little-known deep sea animals. Some of the methods now being used are so crude as to be almost ludicrous and are wasteful of time and money. As one example, a biologist in a submarine collected a number of interesting specimens of sedentary organisms—that is organisms that don't move around. He placed them in a wire basket attached to the sub and brought them back to the surface. As the sub was being lifted from the water it bumped against the mother ship, knocking the basket loose. All the specimens thereupon spilled out and sank to the bottom.

We completed the traverses without difficulty and saw no unusual bottom features that would make this a poor area for our geotechnical research into the bottom muds, so we went on to the next phase.

Probing the Bottom

OUR FINAL TASK ON THIS DIVE WAS TO TEST THE OPERATION OF the new geotechnical probes. The most important unanswered question was this. Would we be able to push the probes into the bottom muds to any significant depth? On the land, a metal rod can be forced into the soil by applying the weight of an object against it. When you push a spade into dirt with your foot, you apply the weight of your body against it. If you didn't weigh anything and had nothing to push against, you would be unable to push the spade into the dirt.

This was somewhat our position in the submersible. The sub was designed to weigh almost nothing when submerged in sea water. It displaced a volume of sea water that weighed almost exactly what the sub weighed in air. When the pilot wanted the sub to float at the surface, he could decrease its effective weight by pumping water out of ballast tanks and replacing that water with lighter air. To make the sub sink during a dive the pilot let water back into the ballast tanks, thus reducing buoyancy. These are relatively. minor changes in the effective weight of the sub.

It is fairly easy, though drastic, to make major changes in the sub's effective weight so as to increase its buoyancy. As a safety feature, the sub can be forced to ascend by discarding or dropping off heavy objects or materials. There are several methods of doing this, all of which have back-up manual procedures that can be followed in case of electrical power failure.

The primary weight-reducing system is shot ballast, composed of steel pellets like BB's or shotgun shell pellets. There are three tanks containing a total of 2,310 pounds (1,650 pounds wet weight) of steel shot that can be released in small amounts or all at once. This system is used to counterbalance the weight of samples or cores that the sub will carry back to the surface. In the ordinary course of events, only enough shot is dropped to provide the buoyancy lift to carry the added weight. However, in the event of a real emergency such as total loss of both main and emergency electrical power, the entire amount of shot ballast is automatically jettisoned.

If more buoyancy is needed in an emergency, the liquid mercury in the trim and list systems can also be jettisoned—pumped overboard—giving approximately 2,200 pounds of additional buoyancy. In even more dire circumstances, another weight reduction of 3,000 pounds can be achieved by dropping the forward main lead-acid battery. Finally, the two manipulator arms and the bow instrument rack can be released. This is a safety feature to be used primarily in case these systems become entangled with objects on the bottom. In total, some 7,500 pounds of additional buoyancy can be achieved.

In contrast to this significant control over positive buoyancy, a major increase in negative buoyancy is not possible during a dive with the current systems and components of the sub. Engineers had calculated that 800 pounds dead weight of negative buoyancy would be the maximum reaction force available for pushing the probes into the bottom.

We switched on the electrical and electronic systems that operated the probes and recorded their output signals. Everything checked out; everything was in working order. The pilot sat the sub down onto the bottom at a location that had previously been selected by the geotechnical engineers. He then filled all ballast tanks with sea water to give the sub maximum weight. The copilot-navigator carefully recorded our position by taking sonar triangulation sights on major sonar targets, such as the concrete blocks and large pieces of immovable bottom junk.

It is ironic that after complaining about the amount of refuse on the sea floor, we actually used it for our own advantage. If the junk had not been there, we would have had to place additional markers on the bottom. And since our markers could not be easily removed, they too would eventually become junk on the bottom. However, the amount we would contribute in one small area would be minuscule in comparison to the junk scattered over thousands and thousands of square miles of sea floor.

For the moment, the pilot and copilot had little to do but watch the geotechnical engineers as they performed the experiments that were the objective of the dive. The electric motor chain drive for the vane shear probe was turned on briefly, and the probe was slowly pushed into the mud a short distance, then stopped. Another electric motor was switched on to rotate the vanes slowly through 360 degrees (one complete turn) in one direction. The amount of force needed to turn the probe for this undisturbed measurement of vane shear strength was recorded on a strip chart recorder. Next the vanes were rotated in the opposite direction for another full turn, and the remolded or disturbed shear strength measurement was made and recorded. Then the probe was pushed down another short distance, and the measurements were repeated.

We repeated this procedure several times until the probe was about a meter deep into the mud. On the next push the observer at the viewport called out excitedly, "Liftoff!" At the same instant we felt the whole sub lift off the bottom momentarily. As we expected, the bottom muds had become stiffer or more resistant to penetration as we probed deeper, and finally, instead of pushing the probe in, we had lifted the sub off the bottom a few centimeters.

The sub slowly sank back to the bottom again under its own weight or negative buoyancy, and we were able to take another set of readings. Obviously we were near our limit at this locality, but we tried one more tactic. As we pushed the probe down with its chain drive mechanism, the pilot attempted to add effective weight by turning on the vertical thrusters to maximum power with the propellers aimed to push us down. It worked. The probe went down another increment, and we took another set of readings.

We might have been able to continue deeper in this fashion, but now a final unanswered question began to worry us. Could we pull the probe back out of the mud by reversing the probe motors? Although no one talked about it, we knew that if we couldn't pull the probe out, we'd have to detach it from the sub and leave it on the bottom—and that would be an embarrassing end to the whole research project. Such a thing was of course unlikely to happen, as we had designed the probes with electric drive motors considerably more powerful than necessary to do their job.

Although the downward thrust was limited by the weight of the sub in sea water, there was no such limit on the upward pulling force. In pulling out, the reaction consists of the whole bottom of the sub pushing against the sea floor. Although the sea floor is soft, it does provide a virtually infinite reaction force against a large surface area. The probe retracted smoothly on the first try. I think that secretly each of us breathed a sigh of relief at not having any trouble with this phase of the work. Still, no one said anything more than "Well, that was easy."

The time was now 1600 hours (4 P.M.) and we had been on the bottom for six hours. We were scheduled to begin our ascent soon. At this point we held a brief discussion in the sub and decided to request permission to stay down a little longer so that we could do two more probes. Our justification was that we had lost two hours in searching for the missing transponder, and now we were just beginning the productive phase of the dive. The pilot called the *Transquest*, explained our request and our reasons to the dive master, and received permission to stay on the bottom for another two hours. The pilot then moved the sub a hundred meters or so to our next probe location and we continued as before.

We completed the second probe without incident and moved on to the third. The operations were by now fairly routine. Imagine our surprise when in the middle of the third probe, all the lights in the sub flickered and went out, and all the electrical equipment suddenly stopped operating. We had lost electrical power. We had experienced a blackout.

There was a moment of utter silence. All the background noises of electric motors, fans, and pumps were gone. For an instant it was frightening as we sat there in the dark cold sub at the bottom of the sea. All sorts of wild thoughts raced through my mind. No doubt we all had our own fears, but I was surprised at how outwardly calm and cool everyone else seemed. Perhaps I seemed calm and cool to them too. Certainly none of us gave voice to our thoughts.

In a matter of seconds, the dim emergency interior lights came on. They operated from separate dry cell batteries and were switched on automatically when the main power cut out. We three scientific passengers stayed quietly where we were and watched the pilot and navigator as they flicked switches and peered at their instruments. Nobody asked "What happened?" or "What's wrong?" or "Is it serious?" We looked at each other, grinned weakly, and said nothing. I wished I had paid more attention to the description of all the safety features.

As long as the pilot didn't appear worried, I wasn't too worried. He had lots of experience and would know what to do. Although I wasn't thinking it out then, it occurred to me later how much wisdom as well as truth there was in the old saw. Panic and fright are contagious and it was important to keep from appearing upset. If one person had said the wrong thing it might have released the fears in all of us. So we said nothing.

Muttering something about checking the fuses, the pilot left his seat and moved into the rear sphere. I had to squeeze aside to give him room to pass through the connecting hatchway, and then I stepped into the forward sphere. As the pilot moved around, the other two passengers had to shift around with him. At the pilot's suggestion, we switched off all the electrical scientific equipment.

In the main power supply fuse box the pilot found that a

circuit breaker and had tripped off. When he reset it, the lights came back on and the fans started circulating air again. There was no obvious indication as to why the circuit breaker had tripped—no wisps of smoke curling out of the instrument rack, no smell of scorched insulation. The pilot continued to move around the sub, feeling, sniffing, looking, but finding nothing.

Moving back to his seat, the pilot called the dive master to report our temporary loss of power, say that the power was now restored, and explain that he could find no obvious indication of the trouble. I think we were all somewhat relieved when the dive master ordered us to return to the surface immediately. The pilot agreed, and we retracted the probe and shut down the scientific apparatus.

We never knew what had caused the power failure. All our scientific equipment operated off the main batteries, and when everything was on at once, we recalled noticing some fluctuations in the voltage. Although it was not exactly an overload, one voltage fluctuation must have been large enough to trip the circuit breaker. Later the electronics engineer made some circuit and wiring changes, thus eliminating the voltage fluctuations.

Ascent and Recovery

BEFORE WE COULD ACTUALLY START THE ASCENT, THERE WERE A number of housekeeping chores to be done. The sub was in a state of negative buoyancy for the bottom probing. To regain positive buoyancy the main ballast tanks had to be emptied. This was accomplished by blowing them out with compressed air that was stored in steel cylinders under high pressure outside the pressure spheres.

A nose-up attitude of the sub was desirable for the ascent. To achieve this, the skipper pumped liquid mercury from a tank near the bow to one near the stern. This did not change the overall weight of the sub and thus did not alter the buoyancy, but it did make the stern sit lower and the nose rise higher. As a final trim measure, the pilot checked the sideways balance. Keeping his eye on the position of a bubble in an inclinometer, he pumped mercury from the port list tank to the starboard list tank until the bubble was centered under the zero list mark.

We all settled back while the pilot and navigator made their final instrument check. Ascending is inherently more comfortable than descending because leaning back feels more secure than tilting forward. The pilot turned on power to the main thrusters, adjusted the stern planes to give us lift, and began to "fly" the sub back to the surface. We couldn't feel any motion, but we could hear a faint hum from the thruster motors. Like the descent earlier that day—it seemed more like days ago than hours ago—the journey to the surface took about an hour. We went up in a helical spiral to stay in approximately the same geographic position.

Until this point in the dive, both the pilot and the navigator had been fully occupied with their duties and had had very little to say to the three scientific passengers. Now that we were safely on our way back to the surface, the navigator—perhaps intending to reassure us—told us about an incident on a previous dive.

"I didn't want to scare anybody by saying anything back there when we lost the electrical power, but it reminded me of that incident nearly a year ago," he began. Then he paused.

"Which incident was that?" the pilot asked, when no one else encouraged the navigator to continue.

"That's right, you weren't on that dive, were you? When we really lost power and aborted the dive."

"What do you mean—aborted the dive?" I asked, my curiosity now thoroughly aroused.

The navigator twisted around in his seat to face us.

"It's sort of a long story," he continued. "We were up at the north end of the Trough last November, doing a dive for the navy. We'd been down a couple of hours, about as deep as we were today. Can't tell you what we were looking for. That's classified. But the power went off suddenly, the same as it did today, only worse." He turned away for a moment to check his instruments.

"Oh yeah, I heard about that," the pilot said. "I was on vacation that month. I've meant to ask you. Was there any warning?"

"We heard a zap and a crackle, then the lights went out and the power was off. But there was still some light, flickering in the back. We turned to look, and there was a fire up there," he said, pointing to a metal connection box above our heads in the rear sphere.

"Sparks or flames?" the pilot asked.

"Both," replied the navigator. "Sparks behind the box, then smoke coming out the front, and insulation beginning to burn at the side."

"Who was with you? Any passengers?" the pilot wanted to know.

"One navy commander, and he was turning a little green back there. Tom was in this seat and I was in yours."

"What did you do?"

"I shut off all the power at the front panel, but a relay must have fused, because that didn't stop it. As long as the sparks were flying, the juice was still on, and a fire extinguisher wouldn't help. And I didn't want any more carbon dioxide in this confined space, anyway. I could think of only one way to cut the juice, and we had to get back to the top in a hurry, anyway." Again he paused to check the sonar.

The suspense was too much for the geotechnical engineer.

"What did you do?" he asked urgently.

"I yelled, hang onto something solid, we're going up in a hurry! Then I dumped the forward batteries."

"Jesus!" the chorus murmured.

"That stopped the arcing, and we felt a jolt as the batteries dropped free. I hadn't had time to tell the TQ, and I just hoped they'd spot us up there. We rocked around a little on the way up, but with the power gone, we couldn't tell how fast we were going. None of the gauges or depth meters were working. I told everyone to hang on tight—it might be rough when we bobbed to the surface."

"How long did it take?"

"Must have been 20 or 30 minutes. I forgot to look at my watch when we started. Then we were busy getting out the face masks to breathe from the emergency air supply. I remembered to tell everybody that we could tell when we got close to the surface because we'd be able to see light out the porthole. About then the pitch-black viewport began to lighten up a little. I watched it a few seconds to see how fast it was getting brighter, and then said, O.K., hang on, we're at the top! I braced myself against the overhead and waited. Then we burst free."

"What did that feel like?"

"It happened so fast there wasn't much to feel. There was an instant of no gravity, like going over the top of a roller coaster. I think there was a quick reverse. First I pressed down into the seat, then lifted out of it as I braced against the ceiling. Then we could feel the rocking of the waves."

"One of the guys on the ship who happened to be looking our direction told me later that we almost popped clear out of the water, only a couple of hundred yards from the ship. They hadn't known what had happened, but they saw us rising on the sonar, and couldn't contact us on the phone. Then they set lookouts in all directions to spot us when we reached the surface. They had the boats out to us in minutes.

"As soon as we felt the wave action and the rocking that the sub does only when it's floating at the surface, I crawled into the after sphere and started undogging the hatch. I wanted out of there fast!"

"Did you ever find out what went wrong?" we asked.

"No, just that some part failed and short-circuited. They had to rewire that system and put in new batteries. We were out of action for several weeks. The old batteries are still on the bottom."

The navigator turned back to his instruments. The rest of us were silent for a few minutes, thinking about that earlier ride to the top. It was somewhat reassuring to hear that the sub could surface in a hurry when there was an urgent reason to do so.

"It's nice to know the emergency system works when it's needed," said one engineer-observer, "but why did you drop the batteries instead of using one of the other less drastic methods?"

"When I saw I couldn't shut off the electrical power, the only way to stop the arcing that fed the electrical fire was to jettison the batteries. There wasn't much time to think, but it didn't seem necessary to drop anything else, such as the shot ballast or the mercury. Dropping the batteries gave us enough buoyancy to surface in a hurry."

At that point, the pilot told us we were nearing the surface. On our ascent the instruments were still working and we knew our depth. We could see some light at the viewport, but it was just at sunset when we surfaced, and there was not much light anywhere. The transition was obvious when the waves set the sub rocking gently. The pilot notified the *Transquest* by radio that we had arrived, and the navigator turned on a flashing strobe light atop the conning tower to indicate our position.

The pilot opened the main access hatch and we all climbed out onto the topsides. It was a real pleasure to stand up straight and stretch our cramped leg and back muscles. It also felt good to breathe fresh air, although the air in the sub never became noticeably stale. There was a psychological effect of five people in a small space for almost 10 hours that made us think the atmosphere was close.

We could see the brightly lit *Transquest* about a half mile away. The balmy evening breeze was comfortable. We relaxed while waiting for the Boston Whaler to come out from the ship. The time was 1900 hours (7 P.M.)—exactly 10 hours since we started the dive.

We waited. And waited. And waited. Nothing seemed to be happening. The night got darker and the breeze turned cooler. The *Transquest* knew we had surfaced; our first surface radio transmission had been acknowledged. Radio contact had then been broken off, and the pilot could get no answer when he tried to ask the reason for the delay.

Gradually the delay began to irritate us. From all that we could tell, conditions were ideal for the pickup and recovery of the sub. The seas were relatively calm, the visibility was good, the wind was light. But the *Transquest* remained dead in the water, silent, floodlights blazing like a casino gambling ship. We were tired and hungry. As the time went by, the wind picked up speed, the unstable *Deep Quest* wallowed, the increasing waves splashed us with sea water that was not quite warm to start with and quickly cooled in the breeze. Soon we were wet and chilled. I think our irritation was at not knowing why there was a delay, rather than at the delay itself. We watched the stars come out, and we griped about our discomfort.

Finally, after nearly an hour of waiting, we heard the popping of an outboard motor and saw a light move away from the side of the ship. Soon the Boston Whaler approached us and the two scuba divers splashed backward over the side, their usual manner of entry into the water. Because it was now dark they carried waterproof flashlights, and we could follow their progress as they swam under water to the sub and performed their tasks of securing the sub in preparation for the recovery.

Now we could see a change in the lights on the *Transquest* as she slowly turned and began to back toward us, maneuvering into a position upwind of us to shield us from wind and currents. The pilot went back into the sub, brought up the remote control box, closed the hatch, and turned on the sub's drive motors. Under power he drove the sub slowly toward the stern well of the ship. One of the scuba divers climbed aboard the sub and crouched, waiting, near the bow.

The floodlights from the ship, now less than a hundred feet away, made the scene as bright as day. A man at the stern shouted and tossed a coil of light rope, called a heaving line, toward us. It snaked out in an arc, carried by the weight of the monkey's fist knot at its end, and fell across the deck of the sub. The scuba diver picked up the heaving line and pulled it in rapidly. This light line was attached to a thicker and heavier line that would secure the ship to the sub. When the loop in the end of the heavy line reached him, the scuba diver crawled forward and secured the loop to a fitting in the bow of the sub. When he was finished he raised his arm in a signal, and then the man on the ship who was tending that line pulled it taut and looped it twice around a power capstan. A capstan is a short vertical metal cylinder that is rotated by a motor. When the line handler pulls his loop tight, the capstan provides the power to pull in the line. When the handler slacks off his loops, the capstan continues to turn but without pulling the line. This gives the handler control without endlessly starting and stopping the drive motor.

Another heaving line fell across the sub from the other side of the stern well, and its mooring line was soon made fast to the bow of the sub. These two mooring lines, kept taut by

Deep Quest *moves in close during the recovery operation after the dive is over. Waves have pushed sub off center. Seconds later it nudges* Transquest, *causing minor dents in the sub's outer hull.*

their handlers, guided the bow of the sub into the stern well. Two more mooring lines were now attached to the stern of the sub.

The crucial point in this operation comes just as the bow of the sub enters the quiet protected water between the arms of the stern well. At this point the stern of the 40 foot long sub is still in the unprotected choppier waters beyond the ship. The sub reacts quickly to waves, whereas the larger ship responds slowly, and only to the larger waves and swells. Because of this, the two vessels move erratically, now in the same direction, now in different directions.

The mooring lines are meant to hold the sub in the center of the stern well, but in this position the stern mooring lines are at too sharp an angle to be very effective. In spite of all precautions a wave caught the sub and carried it against the starboard end of the ship and we felt a gentle bump. Before another wave could catch us, the sub's forward thrust carried us farther into the quiet water and the mooring lines pulled us to the center of the well.

The pilot shut off the sub's motors, and the line handlers tugged and pulled us into the properly aligned positions. At last the dive master activated the elevator platform. The sub was lifted out of the water. The dive was over. We climbed down to the deck of the *Transquest*.

We were told that the cook was saving supper for us, so we went right off to the galley. The cook was in a bad mood, growling at us for making him work late. When we protested that we had been kept waiting for an hour after surfacing, we finally learned the reason for the delay in the recovery of the sub. Because our dive had been extended beyond the originally scheduled time, we surfaced just when the recovery crew had sat down to their evening meal, and they did not come to fetch us until they had finished the meal.

The cook dished up our suppers and left. Soon we were feeling much better. To make up for all the inevitable and unavoidable discomforts of life aboard a working ship, the food on an oceanographic research vessel is always excellent, plentiful, and well prepared. We thoroughly enjoyed the swiss steak, mashed potatoes with gravy, and green beans, followed by fresh-baked cherry pie and ice cream.

As we dawdled over our coffee, I began to speculate about my next move. If possible I wanted to get back to San Diego, as I was not scheduled for any more dives in this series. My replacement was due to arrive at San Diego airport late that night, and some arrangement must have been made to get him out to the ship.

A Wild Journey to Shore

AFTER SUPPER I CAME OUT ON DECK AGAIN TO LOOK AT THE DENT in the outer aluminum hull of *Deep Quest* that had been made during our recovery. The dent was hardly noticeable and didn't appear to affect any working parts of the sub. I wondered who I should see about the schedule for getting back to San Diego. The sub's pilot, Larry, was talking to the dive master, and I went forward to join them. They were talking about sending a boat back to the Lockheed dock for some spare parts, and I volunteered to go along. As the boat was to leave soon, I went up to my bunk to get my suitcase.

While we were eating our late supper, Lockheed engineers and technicians had quickly checked over the *Deep Quest*. Several things had been found to need attention and repairs before the next dive. The electrical failure—our temporary blackout—was traced to a malfunction in the DC to AC inverter, and spare parts for that were not on board ship.

Also, a short piece of nylon rope was found fouled around one of the main thruster propellers. The rope, unlike any used on the ship, had apparently been entangled during the dive. The fouling rope would have increased the drag on the thruster motor and possibly have damaged it. The dive master decided to replace this motor, but the spare was ashore in the shop.

There would probably be no dive the next day, as our long dive today had depleted the sub's batteries and they would have to be recharged for longer than usual. This would allow time for the repairs to be made, if the parts could be obtained without too much delay. The dive master decided to send one of the Boston Whalers on the 20-mile trip back to San Diego. The *Transquest* would remain at sea. Because of its relatively slow speed, the mother ship customarily remained at sea if a delay lasted only a day or so.

When I got back to the main deck with my suitcase, I found that Larry had volunteered to take the boat ashore. This was to be a long day for him, as he had just piloted the sub for 10 hours, but long hours are not unusual on a research ship. We climbed down a rope ladder into the waiting Boston Whaler. A crew member on the ship handed down a couple of packages and my suitcase wrapped in a sheet of plastic to keep it dry. We shoved off from *Transquest* shortly after 2100 hours (9 P.M.) on a clear moonlit night with relatively calm seas.

The twin outboard engines were capable of pushing the Boston Whaler at speeds of 15 to 20 knots in calm water. I tucked my suitcase under the stern seat, held a flashlight in one hand to serve as the required white stern light on a motor vessel at sea at night, and held onto the gunwale of the boat with my other hand. Larry sat forward, handled the wheel and throttles with one hand, and held in his other hand another flashlight with a divided red and green lens to serve as our port and starboard running lights.

Although everyone seemed to take it matter-of-factly, I began to wonder what I had let myself in for. We were in a small outboard motorboat at night, 20 miles at sea, completely out of sight of land. And the boat obviously wasn't equipped for this sort of operation. I watched over the stern as the lights of *Transquest* rapidly faded away. Soon there wasn't a light to be seen anywhere on the horizon. Larry asked me to shine my flashlight on the compass for a moment to set our course, and thereafter he guided us by the stars.

After a while we could see a glow on the horizon ahead of us, and in about half an hour we could begin to make out the lights and skyline of San Diego. The sea was fairly calm, but our bow rode up and down over the small waves and Larry had to throttle down to about 12 knots for comfort.

As we approached San Diego Harbor, the seas picked up a little and we had to turn about 45 degrees to make the entrance to the harbor. We were now meeting the waves at about a 45 degree angle off our starboard bow, and Larry further cut back our speed to prevent excessive slamming of the boat and to reduce the spray in our faces. The ride was much rougher, but we had only a short distance to go to reach the protected waters inside the breakwater.

The tide was running out in a swift current, and at our reduced speed we made very slow headway into the harbor entrance. As soon as we were past the outer end of the stone wall of the breakwater the sea became much calmer. The out-going tidal currents appeared to increase in velocity, however, so Larry opened up the throttle to increase our speed. Surprisingly, we didn't seem to gain much headway. In fact, by sighting on lighted buoys and lights at the end of the breakwater, we learned that we were going backward, not forward. It was obvious that with engines we weren't going to make it into the harbor!

Larry shouted back to me that something was wrong, maybe we'd sheared a pin. (The shear pin holds the propeller to the shaft. When the pin breaks off, the propeller doesn't turn as the shaft turns.) I had visions of drifting back out to sea and of spending the rest of the night wallowing around in a powerless boat. It would be daybreak before we could expect to be spotted by the many fishing boats going in and out of San Diego.

Larry didn't give up so easily. Using his skill and experience with small boats, he angled the bow against the tidal current and worked the boat out of the mainstream but closer to the rocky breakwater. As we drifted backward he spotted a sandbar near the end of the breakwater and somehow managed to beach the boat. Just before we grounded he tilted up the outboard engines so the propellers wouldn't drag on the bottom—and jumped out to pull us ashore. I hopped out too, and together we pulled the boat firmly onto the sand.

With my flashlight, Larry examined the outboard engines. The propellers were firmly attached, but something else looked wrong. The engines were not lined up parallel in the same direction. Each was turned in a different direction, so that the propellers faced each other. One engine was trying to turn us to the left while the other was trying to turn us to the right. Even at full throttle their thrusts nearly balanced so that the boat didn't turn, but in that position they produced almost no forward thrust.

The trouble was in the steering mechanism. Larry traced the cables from the wheel over several pulleys back to the engines and found a broken fitting. This had released the cable tension, allowing balance springs to swing the motors sideways. The increased strain in the slamming into rougher seas just outside the entrance to the harbor had probably done the damage. Ever resourceful, Larry didn't take long to make a temporary repair with the inevitable piece of baling wire. Within 15 minutes after beaching, we shoved off again.

The remainder of the trip was relatively uneventful. The lights of San Diego were pretty, but confusing in their abundance. But Larry had little difficulty in finding his way through the maze, and eventually he pointed out the Lockheed dock. A few minutes later he shut off the engines and we docked the boat. The time was 2115 hours.

Larry was soon ready to go off on his own business, and we shook hands and said our goodbyes. I found a locker room, changed clothes, and set off to walk the few blocks to the airport terminal. Less than 24 hours ago I had walked these same blocks in the other direction.

I had not yet had time to sort out all the events of the day I had spent at the bottom of the sea. I wanted to think about what I had seen on my first dive on a research submersible. I wanted to plan what I would do on my next dive, if I had the chance again. And I definitely wanted to dive again. There was much to be learned by direct observation of the sea floor.

Later, I fastened my seat belt on the transcontinental jet airliner that would whisk me back to the East Coast and the ordinary routines of daily life. Then I drifted off to sleep, dreaming I was flying to the West Coast to take part in another dive on the research submarine *Deep Quest*.

Epilogue

AFTER BEING LAID UP FOR SEVERAL MONTHS, *Deep Quest* HAS AGAIN been activated for a navy contract to test the application of a new primary power source for submersibles—a fuel cell of the kind used on spaceships. If the fuel cell proves to be suitable it will eventually be put to use on the deep submergence rescue vehicle, more commonly called DSRV.

The versatile and abundantly instrumented *Deep Quest* was a necessary step in the development of radically new vehicles and new concepts. It had been designed and built to test features that later went into the DSRV. This new craft is capable of rescuing up to 24 men at a time from a disabled submarine on the sea floor at depths to 5,000 feet.

Similar features went into the design for the deep submergence search vehicle (DSSV), which will be capable of descending to 20,000 feet and recovering small objects. Experience gained from all these projects has led to the development of a subsea oil well completion system, for oil wells on the continental shelf and continental slope that are too deep for conventional techniques.

Transquest is being used nearly full time as a mother ship for testing and evaluating of the DSRV for the U.S. Navy.

Several other research submersibles were built at about the same time as *Deep Quest*, and some of these are still in operation. However, many second generation submersibles have been built more recently, designed for specific purposes and incorporating advanced technology. Most of these are intended for work on continental shelves, at depths of less than

1,000 feet, and are thus less expensive to build and to operate than the versatile and deep-diving *Deep Quest*. These craft, called workboat submersibles, are being used widely by oil companies for underwater oil exploration, servicing of sea floor oil well installations, and inspection and repair of sea floor pipelines.

Despite the apparent emphasis in this narrative on things that didn't work and things that went wrong, a lot was learned and a lot was accomplished on this first dive of the Lehigh–Lockheed Sea Grant series. Thanks in large part to experience gained on this dive, several features of the equipment were improved and several minor flaws in the plan of operation were corrected. Later dives went much more smoothly, and all the objectives of the series of dives were accomplished.

The Lehigh University Sea Grant research project was completed in 1974 after running for four years. More than 20 scientific and engineering research papers have been published on the results.

Many people were involved with the project. The principal investigator, Professor Adrian Richards, is director of the Marine Geotechnical Laboratory and I am director of the Center for Marine and Environmental Studies at Lehigh University. Two other Lehigh professors were active on the project: Dr. Terence Hirst and Dr. Theodore Terry.

Larry Shumaker, the *Deep Quest* pilot on the dive described in this narrative, is now pilot of the research vessel *Alvin* for the Woods Hole Oceanographic Institution. Don Saner, the copilot–navigator, is on the DSRV project. The Lockheed marine geotechnical engineer, Tony Inderbitzen, was granted his Ph.D. degree at the University of California–San Diego and is now manager of oceanographic research vessels at the University of Delaware.

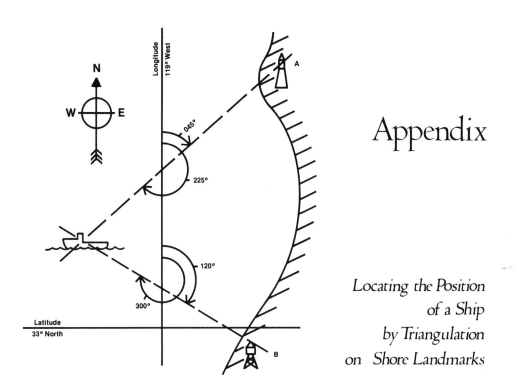

Appendix

Locating the Position
of a Ship
by Triangulation
on Shore Landmarks

A BEARING IS THE ANGLE MEASURED CLOCKWISE FROM NORTH TO the line of sight from the ship to a landmark, such as a lighthouse or water tower, that is shown on a map or chart of the area (see sketch). This line of sight is transferred to the map or chart by drawing it from the landmark toward the ship—a reverse bearing, which is 180 degrees from the forward bearing. A complete circle is 360°, and 180° is one-half of a circle. If the compass bearing from the ship to Lighthouse A is 45 degrees east of north (45°), the reverse bearing is 45° + 180° = 225°. A pencil line is drawn on the chart from Lighthouse A in a direction measured with a protractor 225° clockwise from north, as if the bearing were from the lighthouse toward the ship. The ship's location is somewhere along this line.

Then another compass bearing is taken on another landmark such as Watertower B. The bearing from the ship is 120°, and the reverse bearing of 300° (120° + 180°) is plotted from the watertower on the chart. The intersection of this line with the line from the lighthouse is the location of the ship on the chart. This method is called triangulation, as the ship and the two landmarks form a triangle. This same principle of triangulation was used in a slightly different way for the navigation and position-tracking of the submarine.

In our example, the ship's position is plotted on the chart as slightly north of 33° north latitude and slightly west of 119° west longitude. More accurate measurements are made by dividing each degree into 60 minutes (60′) and each minute into 60 seconds (60″). Thus, the actual position of our ship is 33°4′40″ north latitude, 119°5′30″ west longitude.

In conditions of poor visibility such as fog, somewhat less accurate radar bearings on major landmarks can be used in the same way to find the approximate position of the ship. Visual bearings can be accurate to less than half a degree, whereas radar bearings that are accurate to one degree are considered quite good. Obviously, the less accurate the bearings, the less accurately the position can be determined.

For greater accuracy a third bearing can be taken on another landmark, preferably about midway between the first two landmarks. The plotted lines of the three bearings will almost never exactly intersect on a point but will form a small triangle, called a triangle of error. The best estimate of the ship's position is the center of this small triangle.

The latitude and longitude of our bottom marker had been determined and plotted on the chart of a previous cruise some months earlier. On this trip we wanted to place the ship at that same location. The simplest way to achieve this was to move the ship until it crossed the desired bearing line to one of the landmarks, and then follow that bearing line out to sea until the second bearing was the correct value. For example, if we were to start near the upper edge of the chart in the sketch, we would proceed south until the bearing to Lighthouse A read 45° on the sighting compass. Then we would turn the ship to a course of 225° (southwest) and run down that bearing, taking frequent checks on the bearing to Watertower B until it read 120°. If we were still on the correct bearing to Lighthouse A, we would have reached our destination. If one or both bearings were slightly off, the captain would know from experience which way to move the ship so as to improve the readings. The process would be one of trial and error until the desired position was achieved.

Glossary

Acoustic—related to sound or to a device actuated by sound waves.

Aft—toward the rear or stern of a ship.

Alcyonarian—a group of sea animals related to corals but without a stony skeleton; a subclass or order of anthozoans including sea pens, sea whips, and sea fans.

Anthozoan—a class of phylum Coelenterata that includes alcyonarians, anemones, and corals.

Athwartship—crossways on a ship; from side to side.

Backscatter—the scattering or reflection of light back toward the source from particles suspended in the air or water through which the light is passing.

Ballast—a heavy substance used in a ship to improve stability, balance, or trim.

Bends—the painful and sometimes fatal condition produced when a diver breathing highly pressurized air moves into a lower pressure environment too rapidly. Gas in solution in his bloodstream forms bubbles that may collect at the joints or in the capillaries in the brain, causing great pain and possibly death. The pain in the joints leads to a "bent over" posture, giving this condition its informal name of "the bends." The cure is immediate repressurization, either by descending deeper in the water and coming up more gradually or by decompressing over several hours in a special decompression chamber on board ship or in a hospital.

Buffer—substances in solution in sea water that neutralize both acids and bases so as to maintain the original degree of acidity or alkalinity of the sea water.

Cathode-ray screen—a fluorescent screen on which a high-speed stream of electrons from a vacuum tube is focused to produce a luminous spot; similar in principle to a television screen.

Chain drive—a method of transferring power from two widely separated gears by means of a chain belt, as on a bicycle.

Core—a sample of sea floor sediment, usually obtained in a stiff plastic tube several inches in diameter and many feet long.

Degrees, minutes, and seconds—as used in latitude and longitude, these are measures of distance. Latitude is measured north or south of the Equator, longitude is measured east or west of the Prime Meridian, an imaginary north-south line that passes through Greenwich, England. There are 360 degrees (360°) in a full circle, 60 minutes (60') in a degree, and 60 seconds (60") in a minute. The distance equivalent of 1 degree is found by dividing the circumference of the earth at the Equator (24,818 miles) by 360. Thus, one degree equals about 68.9 miles at the Equator. However, these miles are land miles (statute miles), equal to 5,280 feet. Nautical miles are longer (6,076 feet) and are more convenient, since one degree at the Equator is almost exactly 60 nautical miles, and 1 minute equals 1 nautical mile.

There is a more technical definition of a nautical mile: that it is equal to one minute of arc ($\frac{1}{60}$ of $\frac{1}{360}$) along a "Great Circle" of the earth. A "Great Circle" is formed on the surface of the earth by the intersection of a plane that passes through the center of the sphere. Thus, all of the north-south meridians of longitude on the earth are Great Circles, and so is the Equator—but the east-west parallels of latitude, unlike the Equator, are not, because the distance around the earth as you move away from the Equator toward the poles gets smaller and smaller.

Draft—the distance from the waterline to the lowest point of a ship, usually the keel; the minimum depth of water in which a specific vessel can float.

Fathometer—an instrument to measure the depth of water by measuring the time it takes for a pulse of high-frequency sound to be sent from the ship to the bottom of the sea and for the echo to bounce back to the ship.

Fore and aft—lengthwise of a ship; from bow to stern (sharp end to blunt end).

Forward—toward the bow or front end of a ship.

Geotechnical—pertaining to the properties or characteristics of a soil or sediment including particle sizes and shapes as well as mineral composition; water content (percentage of dry weight) and chemical composition of substances dissolved in the water within the soil; density (wet weight per cubic foot); and shear strength and weight-bearing capacity.

Gunwale—the upper edge of the sides of a boat or ship, pronounced "gunnel."

Hydraulic—related to water or other liquid in motion.

Hydrostatic—related to liquids at rest or to the pressures they exert.

Limestone—a sedimentary rock composed of the mineral calcite (calcium carbonate, $CaCo_3$) formed from the shells of marine organisms.

Lithified—hardened from a sediment into a rock; for example, shale is lithified mud.

Phytoplankton—microscopic and larger plants that float or are suspended in water.

Pinger—a battery-powered instrument that emits pings or beeps or short pulses of high-frequency sound into the water.

Port—the side of a ship that is on your left when you face forward; the direction away from the ship on the left. Port was formerly called larboard, but starboard and larboard sound too much alike, so the change was made in order to avoid mistakes. Compare *Starboard.*

Precipitated—separated out or crystallized from a solution or suspension in water.

Rack and pinion—a metal bar or rod, usually straight, with teeth cut into one edge that mesh with the teeth of a circular gear, the pinion. Rotation of the pinion gear moves the rack in a straight line (linear motion).

Salinity—a measure of the amount of salt dissolved in water; for example, 35 parts per thousand (ppt or $^0/oo$) salinity is equal to 3.5 percent by weight of salt dissolved in sea water.

Sandstone—a sedimentary rock composed of grains of sand, usually of the mineral quartz, that have been lithified or hardened by the addition of a precipitated cement such as the mineral calcite (calcium carbonate).

Scuba—an acronym, or word derived from the initial letters, for *s*elf-*c*ontained *u*nderwater *b*reathing *a*pparatus.

Sedimentary basin—a deep area of the ocean in which sediment accumulates, surrounded by shallower water.

Sedimentary rock—rock formed from sediment; either fragments of rock transported from their source and deposited in another place, or minerals precipitated from solution, or shells of organisms.

Shale—a sedimentary rock from mud, lithified by the heat and pressure of deep burial under other sediments.

Shear strength—a property or characteristic of a cohesive sediment (one that sticks together) enabling it to resist a shearing force such as the slow rotation of a paddle-shaped device.

Sonar—an apparatus that detects the presence and location of a submerged object by means of sound waves sent to the object and reflected back to the apparatus; similar in principle to radar. The word is an acronym derived from *so*und *n*avigation *a*nd *r*anging.

Starboard—the side of a ship that is on your right when you face forward; the direction away from the ship on the right side. Compare *Port*.

Strobe light—a lamp that produces repeated flashes of very intense light.

Submersible—small research or work submarine, usually civilian in purpose.

Thruster—a device for pushing a boat or providing it with motion forward, backward, or sideways, such as a propeller or water jet.

Transducer—a device that transforms power from one form to another; for example, that converts electrical pulses into sound pulses in water (acting as a loudspeaker), or that converts sound pulses in water back into electrical pulses (acting as a microphone).

Transponder—a transducer or pinger that is triggered into action by a pulse of sound or a coded series of pulses of sound.

Triangulation—a method for finding a ship's position at sea by means of bearings (direction of sight from the ship) to two or more objects (such as landmarks or buoys) whose positions are known.

Trim—position with reference to horizontal.

Trough—an elongated basin.

Underwater habitat—an underwater living chamber where scuba divers can live and work for hours or even days without special breathing apparatus, at air pressures equivalent to the water pressure at the depth at which they are diving, without having to decompress till they return to the surface.

Supplementary Reading List

Anikouchine, W. A., and R. W. Sternberg. *The World Ocean, an introduction to oceanography.* Prentice-Hall, 1973, 338 pp.

Barham, E. G., N.J. Ayer, and R. E. Boyce. "Macrobenthos of the San Diego Trough: photographic census and observations from bathyscaphe *Trieste*," *Deep-Sea Research,* vol. 16, pp. 773-784, 1967.

Cousteau, J.-Y., and James Dugan. *The Living Sea.* Ballantine, 1975.

Critchlow, K. *Into the Hidden Environment: The Oceans.* Viking, 1973.

Gaber, N.H. *Your Future in Oceanography.* Arco, 1974.

Gross, M. G. *Oceanography, a view of the earth.* Prentice-Hall, 1972, 581 pp.

Heezen, B., and C. D. Hollister. *Face of the Deep.* Oxford University Press, 1971.

Heintze, C., *The Bottom of the Sea and Beyond.* Nelson, 1975.

Hersey, J. B. (ed.) *Deep-Sea Photography.* Johns Hopkins Press, 1967, 310 pp.

Horsfield, B., and P. B. Stone. *The Great Ocean Business.* Coward, 1972.

Luard, E., *The Control of the Sea Bed: A New International Issue.* Taplinger, 1974.

Mero, J. L. *Mineral Resources of the Sea.* American Elsevier, 1965.

Nelson, S. B. *Oceanographic Ships Fore and Aft.* Government Printing Office, 1971, 240 pp.

Shepard, F. P., Geologic Oceanography, Crane, Russak, 1977, 240 pp.

Wenk, E., Jr. *The Politics of the Ocean.* University of Washington Press, 1972

Wooster, W. S. *Freedom of Oceanic Research.* Crane, Russak, 1973.

Index